Canals of Canada

Canals of the World

CANALS OF CANADA

by
ROBERT F. LEGGET

DOUGLAS, DAVID & CHARLES: VANCOUVER
DAVID & CHARLES: NEWTON ABBOT

This edition first published in 1976
in Canada by Douglas, David & Charles Ltd.,
1875 Welch Street, North Vancouver, Canada
in Great Britain by David & Charles (Holdings) Ltd.,
Newton Abbot, Devon

ISBN 0-88914-045-6 (Canada)
ISBN 0-7153-7195-9 (Great Britain)

Canadian Shared Cataloguing in Publication Data

Legget, Robert F   1904-
      Canals of Canada

(Canals of the world)

      Includes index.
      ISBN 0-88914-045-6

      1. Canals - Canada - History.   I. Title.
II. Series.
HE899.L439   386'.46'0971

Printed in Hong Kong

Contents

List of Illustrations

LIST OF MAPS

'...the Canadian canals ... of extraordinary extent, and admitting a navigation equalled in no other artificial waters of the world...there are no adventitious circumstances to keep them prominently before the public mind. They are seldom spoken of in the public journals. No calamitous accident leads to their mention. There is no connected history of them...'

(From *The Canadian Canals: Their History and Cost*, by William Kingsford, published in Toronto in 1865)

CHAPTER 1

The Waterways of Canada
An Introduction

Waterways constitute one of the glories of the land that is Canada. Despite the size of the country, about the same as the whole of Europe and second largest in the world, it is possible to cross Canada from coast to coast using natural waterways with only short portages that can readily be walked. It is likewise possible to travel by canoe, with many short portages, from Montreal on the St Lawrence to the shores of the Arctic Ocean in north-west Canada. Alexander Mackenzie was the first white man to make both these transcontinental journeys, reaching the Arctic in 1789 after discovering the great river that now bears his name. He penetrated the mountains of the west up the Peace River in 1793, eventually being the first man to reach the Pacific coast of North America from the Atlantic. These waterways of Canada were the transportation routes of the fur trade for well over two centuries. 'It is no mere accident,' a well known Canadian writer has said, 'that the present Dominion coincides roughly with the fur-trading areas of northern North America.'[1] Interconnecting lakes, rivers and streams have therefore played a significant role in the development of the Canada of today quite apart from providing so many beautiful and exciting vistas on the broad canvas of Canadian scenery.

A glance at even a small map will show how the Gulf of St Lawrence penetrates far from the open sea, the River St Lawrence being tidal beyond Quebec City and navigable by ocean vessels as far as Montreal, about 1,000 miles from the Atlantic. One can travel on this river and the Great Lakes through which it flows for another 1,300 miles, to the very heart of the continent. This one river drains an area of 295,000 square miles, two and a half times the area of the British Isles. One of its tributaries is the Ottawa River, itself a mighty stream, its final course in almost a direct line from Montreal to Georgian Bay on Lake Huron. Convenient portages to Lake Nipissing give a through waterway with the French River all the way from Montreal to Lake Huron and so to the other upper lakes. This 'Ottawa Waterway' was that first used in 1615 by Samuel de Champlain and by most of the French explorers who followed him. For two centuries the Ottawa River route was the gateway to the continent; discoverers of the Mississippi, the Missouri and the mid-west all used it as they steadily pushed back the borders of New France and then of Canada.

From the head of Lake Superior all travellers to the west used one of two converging routes which, with portaging, took them into the Lake of the Woods and so into Lake Winnipeg and the great rivers of the Nelson River system. The Nelson flows from Lake Winnipeg into Hudson Bay. It therefore provided access to the prairies for men of the Hudson's Bay Company and many others who sailed into the Bay. One of its tributaries, the Red River, flows due north from the Dakotas into Lake Winnipeg, providing access to the Canadian west from the United States. The Saskatchewan River, with its two great branches extending all the way to the mountains, flows into the same lake and so provided the waterway not only to the far west but also to the Arctic, as a famous portage joins it with the Clearwater River and so with the basin of the Mackenzie River. With an area of 700,000 square miles, this drainage basin makes the Mackenzie one of the great rivers of the world. The Nelson basin is second to it with an area of 414,000 square miles, three and a half times the size of the British Isles. Short portages between both the Mackenzie and the Saskatchewan watersheds lead to streams flowing into the Pacific. Canoe journeys of heroic character along routes thus naturally provided had opened up most of Canada by the early nineteenth century and provided the essential lines of communication for the far-flung fur trade, first of the French, then of the French and British partners of the North West Company and the men of the Hudson's Bay Company.

The geology of Canada alone accounts for this network of waterways. Practically the whole country was glaciated, the last ice retreating to the north about eleven thousand years ago. In its slow retreat, not only did the ice cover wear down the rocks of the Shield to a reasonably uniform maximum elevation but escaping waters from successive vast glacial lakes formed great spillways that were later occupied by more recent rivers. Many of the key portages of Canada will be found to be along these glacial spillways so that the frequent juxtaposition of separate river systems can be readily explained. The Indians, in the course of the centuries, discovered these good routes and used them for their primitive purposes, showing them to white men, such as Champlain, with whom they established friendly relations. The same geological reason explains why Canada is blessed with so many lakes, lakes innumerable it may well be said, with an estimated total area of 292,000 square miles, or 7.6 per cent of the total area of the country, the greatest area of fresh water lakes in any comparable area of the world. The lakes are frequently enlargements of rivers and streams so that they became vital links in the chain of water routes used for canoe travel.

Partners of the North West Company were well named the 'lords of the lakes and forests'. Their canoe routes extended at their peak almost 3,000 miles from Montreal, putting an almost impossible burden on the fur-trading system. It was, therefore, no surprise to find that partners of the company were the first to investigate and to execute improvements to the waterways which they used to such good effect. In 1797 they constructed the first small lock on the St Mary's River

between Lakes Superior and Huron. They constructed one
of the first trading vessels on Lake Ontario, thus participating in the gradual changeover from the Ottawa to the St
Lawrence as the main commercial route to the west. They
installed tramways at difficult portages, some of which were
eventually changed to small canals. And they saw the beginning of the use of *bateaux* and Durham boats as commercial
successors to the canoe just prior to the revolutionary introduction of steamboats.

The great journeys of the fur-traders from Montreal to the
head of Lake Superior were made in *canots de maître*, the
ultimate development of the Indian birchbark canoe. Generally 36 ft long with a beam of 6 ft, these magnificent
craft—entirely made from products of the forest—would
carry payloads of 3 tons in addition to their crews with their
personal loads, a total load of as much as 4 tons. They were
too large and heavy for the more difficult western routes for
which the *canot du nord* was developed, similar in basic
design but with only half the capacity of the larger canoes,
their length being 25 ft.[2] Any damage to these frail vessels,
containing not a scrap of metal, could be repaired en route,
from birch and spruce trees in the forests. Something
stronger was needed for freight movements over shorter
distances when settlement started along the rivers of eastern
Canada. The common *bateau* was the first answer, a strong
wooden flat-bottomed scow-like craft capable of carrying
from 2 to 4 tons of freight. As greater loads had to be moved,
it was developed into the Durham boat, a similar scow-like
craft also with a flat bottom, which could carry up to ten
times as much as a *bateau*. Small sailing ships of conven-
tional design came to be used on Lake Ontario, but the
advent of the steamship in the early part of the nineteenth
century gradually replaced all such simple vessels.

Canada's first steamship was the *Accommodation*, built in
Montreal and equipped with a Boulton & Watt engine from
the Soho foundry; it made its first trip from Montreal to
Quebec in November 1809.[3] It was little more than a scow

with an engine mounted on it, as were many of the earliest steamboats, but designs soon became more sophisticated, especially after complete steam vessels were brought in from the United States, and even imported from Great Britain in prefabricated form. It was not too long, therefore, before the vessels on Canadian waterways were comparable with those sailing in other countries. In writing of his 1842 visit to Canada, so acute an observer as Charles Dickens recorded that 'The (Canadian) steamboats on the lakes in their convenience, cleanliness and safety; in the gentlemanly character and bearing of their captains, and in the politeness and perfect comfort of their social regulations, are unsurpassed even by the famous Scotch vessels, deservedly so much esteemed at home.'[4]

The Ottawa River had its first steam vessel only in 1820 but, after this slow start, steamships quickly took over almost all freight and passenger movements by water until they, in turn, were almost completely superseded by steam railways. As the west started to open up, steamboats were seen to be essential. Many were taken out to the west in prefabricated form, to be hauled in pieces over portages and even carted across the plains for erection on the banks of the rivers they were to serve. The Red River had such a vessel from the east as early in 1851. This was just the beginning. Lake Winnipeg became well supplied with steamers, some actually being hauled up over the Grand Rapids at the mouth of the Saskatchewan to form the start of the Saskatchewan river fleets. In the difficult country centred on the Lake of the Woods (between Ontario and Manitoba of today) prefabricated steam tugs were hauled in pieces through the forest, portaged over trails and then re-assembled, to form a little fleet for assisting travel over this difficult 400 mile water route prior to the coming of the railways. Farther east, the Ottawa River had a well developed steamship service all the way from Montreal to Ottawa and beyond (with stage coach and other connections around old portages) all the way up to the junction with the Mattawa River, 308 miles from

Montreal. With the penetration of the Ottawa Valley by railways, this service ceased abruptly in 1879 apart only from the service between Ottawa and Montreal. This continued, with a grand fleet of great white ships, until just before World War I, one of the vessels chartered in Ottawa being licensed to carry 1,100 passengers.[5]

Many of these early steamship services utilised the canals of Canada which it is the purpose of this book to describe. There were locally famous steamship companies on both the Rideau and the Trent Canals, for example, as well as on the Great Lakes making use of its canal connections. Although some local freight services by water continued after World War I, all but a few passenger services had gone. In the 1930s, however, one could still enjoy a sail from Toronto across Lake Ontario to Queenston for a visit to Niagara Falls. From Toronto one could still take a splendid old paddlewheeler to Rochester, New York, thence to Kingston and finally to Prescott at the head of the St Lawrence rapids. Here one transferred to a smaller screw vessel, the *Rapids Prince* in the last years of this fine service, and on this fair-sized craft one could proceed to shoot the rapids, enjoying quiet sails in between swift sections of the St Lawrence all the way to Montreal. There were still some splendid cruise ships on the Great Lakes, both Canadian and of US registry, but they, too, gradually disappeared.

One of the last remaining services, operating until the mid-sixties, was that worked on the St Lawrence from Montreal to Quebec as an overnight service. Splendidly equipped vessels, immaculately maintained with gleaming white hulls—and white paint was a tradition for almost all such Canadian passenger vessels—left Quebec and Montreal just before dinner, arriving at their destinations at breakfast time next morning, a thoroughly civilised way of travelling. The eastbound vessel continued down the St Lawrence at summer resorts on the north shore before turning up the Saguenay River, the head of which was reached late in the evening after a sail that justly became world-famous because of the magnificent scenery, fjord-like on a majestic scale. This service, too, was finally abandoned in the winter of 1965, the economics of the essential replacement of the old vessels which provided the service being apparently responsible for this regrettable decision: certificates for their 1966 operation could not be issued.

By strange coincidence, the last sailing of this most famous service was a special charter voyage for those attending the Sixth International Soil Mechanics Conference in Montreal in October 1965. I was at Bagotville (the Saguenay terminus) to greet the boat on arrival, the only non-staff person on the wharf. Little did I dream as I took the photograph to be seen on p 56 that I was the solitary public witness of the end of a transportation era.

But the waterways remain, even though all of the great white fleets of steamships have gone. The St Lawrence has now been developed into a great Seaway, permitting large ocean vessels to sail as far inland as Minnesota. Large diesel tugs push strings of laden barges every summer down the Mackenzie River in a part of the land where yet no railways or roads are to be found. The lakes, the rivers and the streams of southern Canada are likewise busy at the height of summer, but now with the canoes of vacationers and sailboats of amateur sailors who dutifully disdain the ever increasing fleets of motor-operated pleasure craft, increasing to such an extent that traffic jams are not uncommon in July and August on some of the canals of Canada.

These vignettes of Canada's natural waterways have been given pride of place not only as a fitting introduction to this review of man-made waterways but also to explain why it is that canals in Canada are so very different in concept from the canals of Europe, of Great Britain in particular and even, to a large degree, those of the United States. Nature's own waterways have given to Canada those water connections which the more usual type of canal was built to provide. There are, therefore, no winding canals making their own way across level ground between rivers far removed from one another. All the canals of Canada are improvements to navigation on existing natural waterways, either in the form of bypasses around rapids or short cuts between adjacent bodies of water, existing waterways being sometimes improved for even this type of canal. There is only one short stretch of canal in Canada known to me that even 'looks like a canal' (and all canal lovers will know what I mean by that!), this being close to the northern end of the Chambly Canal where the Richelieu River, which it parallels, cannot be seen.

The canals of Canada are distinct in yet another important feature. Although some of the very early canals were started by private companies, one or two being privately operated for relatively short periods, most of the country's artificial waterways were built as public works by governments. Since the time of Confederation (1867) the canals of Canada have been publicly owned, with one minor and somewhat unusual exception in the mountains of British Columbia. The Government of Canada has been and still is responsible for almost all of the country's canals, the only exceptions today being three purely local improvements in the province of Ontario. This remarkable fact suggests that there must have been special historical factors influencing the development of canals in Canada instead of the more usual demands of commerce as in the countries of Europe. There were indeed. It will therefore be desirable to get an overview of the story of Canadian canals before they are considered in more detail, individually and in relation to their settings.

Although there had been a few proposals for short canals in the latter part of the seventeenth century, it was not until the early part of the nineteenth that canal building really started in Canada, with three minor exceptions which we shall review later in their local context. 'Canal fever' came late to the little colonies on the St Lawrence, but this is not surprising when thought be given to the wild state of the country even at that time and the very limited development of permanent settlements. The entire country, apart only from the small clearings along the St Lawrence below

Montreal and a few fledgling settlements upstream, was still virgin forest, trees coming right down to the water's edge along all rivers and streams. Even by the year 1825, the population of Montreal was still only 22,357; that of the village of York (now Toronto) a mere 1,677. There were no roads to speak of outside of the few settlements. All travel was by water in summer and over the snow and ice of winter. The task of canalisation, therefore, when the first canals were authorised, was a very different matter from canal construction in England, for example, since all supplies— apart only from timber, charcoal and (sometimes) lime—had to be laboriously brought up by water or over the ice from Montreal or Quebec. There was no trouble with Indians, who were by then pacified, but very real problems were occasioned by the extreme heat of eastern Canadian summers and the associated swamp fever, with the distressing nuisance of biting flies so bad that it is impossible adequately to describe it.

Down by the sea in the still separated colonies of Nova Scotia and New Brunswick, some small canal projects were proposed and two were started near the middle of the nineteenth century, both short cuts along the lines of well established portages. Travel and trade between settlements on the St Lawrence and New York had long been aided by the natural waterway provided by the Richelieu River, Lake Champlain at its head and the adjacent upper valley of the Hudson River leading to the south. Canalisation of this well-established route was an obvious opportunity for gaining the advantages that through traffic by water would present. Work at the New York end was proposed as early as 1792 but did not start until 1817, the (US) Champlain Canal being opened in 1825. Such was the interest in canals at that time that in the 1820s a canal between Boston, Massachusetts and the St Lawrence was quite seriously proposed, a survey of the suggested route being carried out on foot.[6] Somehow it was to get through the mountains and into the valley of the Connecticut River, thence up and over the

Canadian border into Lake Memphramagog and then, somehow, to the St Lawrence. There were probably other equally visionary ideas, but canalisation of the Richelieu River was the only project carried out as part of a waterway link between Canada and the seaboard United States, and then only in years following 1812-1816.

These dates will have little significance outside of North America, but they are the years during which Great Britain was last at war with the United States of America, battles being fought on land in Canada as well as on US territory, naval battles being fought on the Great Lakes and Lake Champlain as well as along the Atlantic coast of the United States. 'The War of 1812' is its official name; 'Mr. Madison's War' is the more popular description of this really lamentable conflict which should never have started. The final fighting (near New Orleans) took place after a treaty of peace had been signed, an oddity of history in keeping with so many unusual aspects of the war. The final result can best be described as a stalemate or draw with the result that written accounts differ in their assessments of it depending on the country of publication. At the time, it was naturally a very serious matter for both countries and especially for their devoted naval and military leaders. It aroused strong patriotic feelings in many of the early settlers near the international border and these did not disappear with the formal signing of a treaty. Feelings continued to run high for some years thereafter; they led to one of the most remarkable canal projects in North America.

Defence authorities of Britain and the United States were fearful that there would be another outbreak of hostilities, and made plans accordingly. Kingston, the Canadian fortress at the outlet from Lake Ontario, had been a vital naval base during the war, but all supplies had to be brought to it from Montreal up the St Lawrence. This was an international waterway for the last 100 miles of this hazardous journey, then involving portaging around the many rapids that made the great river so beautiful and yet so dangerous.

The Duke of Wellington decided that an alternative route between Montreal and Kingston must be found as a military precaution. Surveys were made; an old Indian route was found connecting the Ottawa River with the St Lawrence near Kingston through two smaller rivers and a chain of lakes. Lieutenant Colonel John By of the Royal Engineers was directed to come to Canada and get this route canalised. This he did between 1826 and 1832, the Rideau Canal being one of the greatest civil engineering works carried out up to that time in North America. Canalisation of rapids on the Ottawa River between Montreal and the entrance to the Rideau Canal had also to be carried out to make this military route complete. These small canals, too, were built at the same time but by the Royal Staff Corps, a little known regiment of the British Army. Used for the movement of military personnel and supplies, but never in warfare, the Ottawa River and Rideau Canals were well used for commercial purposes from the very first years of their operation. They constituted the first 'Seaway' between Montreal and Lake Ontario with access thence to the upper Great Lakes. They served in this way for about two decades, until the St Lawrence River canals had all been completed, thus providing a more direct route from the sea to the Lakes.

The St Lawrence is today such an obvious route up from Montreal to Lake Ontario that it is difficult to understand why the Ottawa River should have been the first to be canalised until it is recalled that at the start of the nineteenth century the Ottawa waterway was still the route used for almost all travel to and from the Great Lakes and the west of Canada. Trading in furs was still the dominant commercial activity and bales of furs of great value could be carried on men's backs up and down the many portages that the Ottawa route involved. Even the later fur-traders saw that this would have to change but it was probably the beginnings of more general trading that swung interest to the use of the St Lawrence route to Lake Ontario. This was confirmed by the construction of the Erie Canal, linking tidewater at New

York by way of the Hudson River, Albany, and the new canal along the Mohawk Valley, with Buffalo and Lake Erie. Officially opened in 1825, the Erie Canal immediately demonstrated the advantages that canals could bring, even though it provided only a 4 ft depth and so could accommodate only scows (or barges).[7] It must have had some influence in ensuring the final start, after much discussion, of the first 'modern' Lachine Canal also completed in 1825. This canal circumvented the great Lachine rapids of the St Lawrence, connecting the harbour of Montreal with its access from the sea and Lake St Louis, the enlargement of the St Lawrence immediately below its confluence with the Ottawa River. Lachine was therefore the starting point for all earlier journeys upriver, whether by way of the St Lawrence or the Ottawa, its name being a constant reminder of the vain hopes of the early explorers that up one of these rivers they would find the way to China and the riches of the East.

A small start had been made in the 1790s by men of the Royal Engineers at building small canals to bypass the next rapids on the St Lawrence just above Lake St Louis. With some rebuilding these had to suffice until 1845 when the first modern canals with masonry locks were completed. Canals around the other rapids that had to be passed before Lake Ontario was reached were completed within the next decade, the whole system providing a depth of 9 ft of water over lock sills. Traffic did not come up to expectations, to some degree because of the well established position of the Erie Canal, even though with its rebuilding it still gave a depth of only 7 ft over sills, its locks being able to accommodate barges of only 240 tons capacity. The total volume of traffic continued to increase, with the gradual opening up of the west, early shipments of wheat pointing the way to what would become a major item in transportation economics.

Discussions, economic and political, continued down the years, with many committees and Commissions making studies in depth of the future of the St Lawrence canals; concurrently railways were becoming well established, posing yet another threat to the further development of water transportation. Despite all the problems and question marks, Canada embarked upon another complete rebuilding of the St Lawrence (and Lachine) canals in the last quarter of the century, a depth of 14 ft being now provided. Tolls were still being charged (until 1903) even though they had been abolished on the Erie Canal in 1883. The rebuilding of the Erie Canal did not start until 1905, and then to limited dimensions. The depth over sills today is only 12 ft, except for the 13 ft deep section between Oswego and the Hudson River.[8]

The fourteen foot St Lawrence canals soon showed their worth, the volume of freight they carried doubling in the first decade of this century, doubling again by 1924 despite the disruption of normal trade caused by World War 1. This result was due in large measure to the corresponding canal development that had been taking place across the Niagara Peninsula. The Welland Canal has always been a Canadian project, even though serving ships of all nations and especially of the United States. It has to provide a passage across the Niagara escarpment with a total rise of 327 ft. The first bold but successful attempt to provide this unusual waterway was a private venture, the first vessels to sail through the initial canal doing so in 1829. Successive rebuildings to the 9 and 14 ft depths were undertaken in concert with the reconstruction of the St Lawrence canals.

There developed, however, a fine fleet of upper-lake vessels much larger than could be accommodated by the 14 ft canals which brought ever increasing volumes of wheat down to the upper end of the Welland Canal, there to be transhipped into lower lake vessels which could then sail directly through to Montreal. Finally, a great decision was made just before the outbreak of World War I to build a fourth Welland Canal with a depth of 30 ft over sills, and locks 859 ft long, capable of passing the largest vessels on the Great Lakes, and even a large proportion of all ocean-going vessels if ever they could get into Lake Ontario. Officially opened in 1932, the fourth Welland Canal is one of the great canals of the world, still in use and now carrying more than 60 million tons of freight in each open season. With improved locks at Sault Ste Marie, the largest lake vessels could now sail from the head of Lake Superior down into Lake Ontario. Until 1959, however, they could go no farther, having to discharge all eastbound cargoes for transhipment to Montreal still in the 'fourteen-footers'.

A treaty had been signed between Canada and the United States as early as 1932 covering the development of power in the international section of the St Lawrence below Lake Ontario and the associated construction of a deep waterway. Opposition in the United States, reflected to a minor degree in Canada, successfully prevented the implementation of the treaty for almost a quarter of a century, but the year 1959 finally saw the official opening of the St Lawrence Seaway as the culmination of the development of the canalisation of this international river, here so briefly summarized but described in more detail in Part Two of this book. Even there, a summary only of the tangled negotiations surrounding the slow progress towards agreement on the construction of the Seaway and associated power development can be given, so involved was it with US political procedures.

It will have been obvious that political considerations had a good deal of influence upon decisions regarding purely Canadian canals, so that Canadian political matters will call for repeated reference in what follows. It may be helpful to readers who are not familiar with the Canadian political scene, therefore, to present an outline of the key events in the history of the modern nation, most of which had some bearing on the use and improvement of its waterways.

The Treaty of Paris of 1763 confirmed the transfer of power for the tiny settlements on the St Lawrence from French to British authority. A second Treaty of Paris in 1783 set the boundaries of the new United States of America, all of the scattered settlements along and up the St Lawrence as far as the Great Lakes being still known as Quebec. The Constitutional Act of 1791 divided this vast area into Lower and Upper Canada (today Quebec and Ontario), the Ottawa River being generally the boundary. For political reasons the two areas were again united in 1841 into the United Province of Canada, an arrangement that probably created more problems than it solved, leading eventually to the passage of the British North America Act by the British Parliament in 1867 and the proclamation of the new federated country of Canada on 1 July of that year. In 1870, and not until that year, the Hudson's Bay Company finally yielded its immense territory to Canada, Manitoba becoming the sixth province. In later years other provinces were formed; in 1949, at its own request, the 'oldest Dominion' of Newfoundland joined Canada as the tenth province.

Prior to 1841, public works were still generally small in size and few in number but the colonies of the time were stirring with the beginnings of their modern development. The United Province of Canada therefore established a Board of Commissioners of Public Works. We shall see how important a role the Commissioners played in early canal development. They had expended about $20 million, a great sum for those days, by 21 December 1867 when the Parliament of the new country passed an Act establishing its Department of Public Works.[9] This Department, still a vital part of the Government of Canada, became responsible for all public works, including railways and canals, and remained so until 1879 when a separate Department of Railways and Canals was established by splitting off from public works these two important branches of transportation. 'Railways and Canals', as the new Department was colloquially known, continued to carry this responsibility until 1935 when it was renamed the Department of Transport. It broadened its responsibilities by absorbing, for example, the Department of Marine, one of the tasks of which had been the development and maintenance of the St Lawrence Ship Channel, the dredged section of the St Lawrence River below Montreal.

'Transport' continued to be responsible for all canals for

the next quarter of a century, through its Canal Services branch. When the St Lawrence Seaway Authority became operational, after its establishment in 1952, it naturally assumed control over the St Lawrence and Welland Canals, later taking over also the Canadian lock at Sault Ste Marie and the Canso lock in Nova Scotia. The Authority, as a public agency, reports to Parliament through the Minister of Transport so that canal administration was still co-ordinated in the one Department. In February 1972, however, responsibility for all canals other than those under the St Lawrence Seaway Authority was transferred to the Conservation Program of the National Historic Sites Branch of the Department of Indian Affairs and Northern Development. This Department includes in its wide range of operations the National Parks Service of Canada with which the older canals are also now associated. There is, therefore, some justification for a move that, at first sight, appears to be somewhat unusual; the Rideau Canal, for example, may well be regarded as a 'national historic site' set in what could be a national park. At the same time, the older canals are still operating canals and they remain vital parts of the transportation system of Canada so that their retention in the Department of Transport would have been equally logical. One strange result of this modern managerial development, further comment upon which would be quite inappropriate, is that the half-mile one-lock St Peter's Canal in Nova Scotia is now linked, managerially, with the canals of Quebec and Ontario, over 1,000 miles away, instead of with the Canso lock, less than 30 miles away but now under a different department.

Enough, however, of administrative matters! The same engineers supervise and direct the excellent major maintenance work on the older canals just as the same devoted lock-keepers and their staffs continue to operate the locks, still taking pride in their local maintenance work and the tending of the gardens that are to be found around almost every lock, despite all the 'changes at Ottawa'. The locks are tended naturally only during the open season of navigation. Since traffic through the older canals is now almost wholly on pleasure bent, it is seasonal in character. Although there are some transits almost from the opening of navigation, traffic is at its peak during the months of July and August. Such peaking of traffic is never easy to handle but the faithful lock staffs manage wonderfully well, especially since most of the locks are still hand-operated. Electrification has been carried out at a few of the older locks, always after a burst of public protest, but it is bound to become more widely essential as traffic continues to increase. If electrification is carried out as carefully as it has been at Newboro lock on the Rideau Canal, for example, the aesthetic and historical nature of the locks will be well conserved while modern convenience is still served.

On the St Lawrence system, however, traffic is steady throughout the entire season. Here the main problems arise at the end of the season when ocean vessels far up the Seaway find time slipping away with the closure of the Seaway in prospect. There has to be a closing date, however, since icing problems make the operation of the locks impossible and navigation hazardous. It is surprising to find that it is possible to keep the Seaway open now for 250 days in the year as compared to a normal operating season of only 222 days when it opened in 1959. The winter periods are not lost, all major maintenance work starting just as soon as the last vessels have passed. With a few ocean vessels now coming up to Montreal throughout the winter, and this despite all problems of ice, what the future holds for the Seaway is still uncertain. There are some who think that, because water at the bottom of channels and locks is at 4°C, it should be possible to circulate this and so keep navigation going on a twelve-month basis. The problem of floating ice, swept downstream by the regular flow of the river wherever it is open throughout the winter, appears to pose an insoluble technical obstacle, but the dream remains. In the meantime, through devoted work on the part of all concerned, the

Seaway continues its vital operation until early in December
of each year, the vagaries of the Canadian climate ultimately
deciding when closure is essential.

All such technical matters affecting the international section of the St Lawrence, as of all other 'boundary waters'
shared by Canada and the United States, come under the
jurisdiction of a little known but highly respected body
which it is a real privilege to mention in concluding this
introduction to our study of the canals of Canada. This is the
International Joint Commission, truly joint since it consists
of three Canadians and three US citizens, with one from
each group acting as joint chairmen. It was established by
the neighbour countries in 1909 and for half a century its
members, then usually a majority of engineers, managed to
settle all disputes about boundary waters with minimum
publicity, perhaps too little. Normal operation of the Commission was interfered with to a degree when the Columbia
River treaty was being negotiated, but the Commission continues its splendid work to the benefit of both countries.
Surveillance of the St Lawrence River and all associated
lakes and waterways continues to be probably the greatest
responsibility of this small body, an example of true international co-operation that is admired around the world by all
who know it.
    From the earliest days, engineers have dreamed how they
might develop the canals of Canada. Some visionary plans
were realised, as we shall see, only to fall into disuse when
the expected traffic failed to materialise. Others were built as
defence measures but became commercial or pleasure-boat
waterways. Today these serve welcome guests from the
United States in place of the military invasions originally
anticipated. Several of the older canals are now well into
their second century of service, with their original masonry
locks, testimony to the sound building of those early days. In
this, as in many other ways, the early development of

Canada owes much, as we shall see, to the good work of the Corps of Royal Engineers of the British Army.

The story of Canada's canals includes even one of the biggest question marks in the history of Canada. For almost a century there was continuing argument between advocates of the Ottawa River as the best route for the Seaway to the Great Lakes and the proponents of the St Lawrence as the more desirable route. The results of the General Election of 1911 decided the issue rather than any detailed economic study but even so it was not until 1959 that the St Lawrence Seaway was finally opened for use by large ocean-going vessels. The steadily increasing volume of freight passing up and down the Seaway clearly demonstrates what potential such a deep waterway has as a part of the transportation system of a continent, a potential of increasing significance in view of the vital necessity of conserving energy in all forms as far ahead as can now be seen. Despite current financial problems, the Seaway is the real success story of Canadian canals but there were also disappointments and some outright failures, all making up the involved but fascinating story of how Canada's canals have supplemented her natural waterways, a story to which this small book can be but an introduction.

PART 1

CHAPTER 2

Short-cuts in the Maritimes

The Maritime Provinces of Canada are well named. The surrounding sea is their waterway and has been since the time of the earliest settlements. A glance at even a small-scale map of this part of Canada will show, however, that there are several narrow necks of land impeding navigation between adjacent inlets from the ocean. Canals across such isthmuses would provide obvious short-cuts; this has provided the pattern for canals in this oldest and quite delightful part of the Dominion. History has its part to play in the recounting of Maritime canal development. It is a turbulent tale but may be summarised with a reminder that, after much fighting on land and sea, the beginnings of the Canada of today were granted to Great Britain by the Treaty of Paris in 1763. Representative (British) government had existed before this in Nova Scotia, from 1758. The first Legislature of New Brunswick met in 1784, that of the smaller province of Prince Edward Island having preceded this in 1769. Cape Breton finally joined with Nova Scotia in 1820. In 1867 Nova Scotia and New Brunswick became two of the founding provinces of the Dominion of Canada, Prince Edward Island joining only in 1873. Although a party to the initial discussions, Newfoundland waited until 1949 before becoming the tenth province as had so long been hoped.

When the four provinces by the sea are referred to, the term Atlantic Provinces is now common usage. Since Newfoundland, however, has not had need of any canal construction, our review will be confined to the Maritime Provinces (the original three) and of these Nova Scotia takes pride of place in this context. A glance at the map will show why this should be so, the province being almost three islands. No point in Nova Scotia is farther than 35 miles from the sea. Its area of 21,425 square miles may usefully be compared with its coast line of 4,625 miles, the ratio of about 4½ square miles of land for every mile of coast being some kind of record, not unrelated to the fact that about one sixth of the total sea coasts of the world are in Canada. A map will show how obvious was a canal across the isthmus of Chignecto, proposals for this canal being the earliest of all in Canada. The canal, however, has not yet been built although the concept still has its enthusiastic supporters. Access into the Bras d'Or lakes of Nova Scotia from the sea to the east was another obvious short-cut. This canal has been built and is still in use after more than a hundred years of service. Not quite so clear from a small scale map is the possibility of linking the harbour of Halifax with Minas Basin at the head of the Bay of Fundy but a canal was also constructed to provide this short-cut, fated to have a very short life of useful service. Finally, and as the reverse of the usual procedure, a short canal and great sea lock have had to be constructed in recent years in order to provide passage through a great rock dam that was built across the Strait of Canso to replace the train ferries that previously provided the only link between Cape Breton and the mainland of Canada.

The Proposed Chignecto Ship Canal

The boundary between Nova Scotia and New Brunswick was logically placed at the narrowest part of the isthmus of Chignecto, the direct distance between Baie Verte to the northeast and Cumberland Basin to the southwest being about 15 miles. The boundary follows the small Missaquash River; the terrain is quite flat. The strategic importance of this location is so obvious that it is not surprising to find remains of Forts Lawrence and Beausejour located close to the boundary, now being restored as historic sites. The low relief and the absence of bedrock near the ground surface suggests that construction of a canal across the isthmus would be a simple task. Unfortunately, the high tidal range in the Bay of Fundy makes this apparently easy short-cut one of complexity both economically and from the engineering point of view. The Fundy tides are world famous and were long regarded as the highest in the world until tidal ranges in the Canadian Arctic that were even greater were found. The more than 50 ft range experienced at the head of the Bay of Fundy made essential the provision of some type of lock at the Bay end of any canal across the isthmus in order to accommodate vessels coming from Northumberland Strait until the period of high tide when they could be floated out and down the Bay. The necessity for this facility became the more pronounced as the vessels to be accommodated increased in size.

This problem of the tidal range was to be a continuing impediment for all the early proponents of the canal. The first of these appears to have been a commissioner specially appointed by the great King Louis XIV of France. Reporting to the 'Sun King' in 1686 Jacques de Meulles said that 'There are only four leagues to be traversed by land to go from French Bay (Sackville) to Baie Verte, and the portage may be easily cut through by canal, since all the land is very low.... If the sea once passed through, it would make in a very short time, a very fine river, through which ships from Quebec could easily pass.'[2] This attraction of the great saving in sailing distance from points on the St Lawrence would continue down the years to be the great argument in favour of the canal project. It would be a saving of about 300 miles between the St Lawrence and all points on the Maine coast and south of this, and of even more for the growing port of Saint John, located on the west side of the Bay of Fundy.

With the start of modern civil engineering, more accurate surveys and estimates of cost became possible. A favourable report, the first of many, was made by Surveyor General

Lockwood in 1819, followed in 1825 by the detailed surveys of an early engineer, Francis Hall, whose estimate for the entire project (including seven locks) was £67,728. Another survey was made in 1838 in which year the Legislature of Nova Scotia passed an Act providing for the formation of the Cumberland Canal Company with a capital of £250,000, but nothing more transpired. In 1842 yet another survey was made by Captain H.O. Crawley RE. He sounded the first note of caution saying, in his report of 19 January 1843, that 'so much uncertainty appears to exist that the project would be extremely hazardous,' advising the lieutenant governor that 'it is not desirable to prosecute the enquiry further.' Incidentally, Captain Crawley states in his report that no less a man than Thomas Telford, the great British civil engineer, had been consulted about Francis Hall's report but no record of Telford's comments has yet been traced.[3]

The enthusiasts were not to be quietened, however, and so the inquiries continued, almost twenty having been made in a formal manner up to 1875. As the development of civil engineering permitted rational solutions to the tidal problem to be suggested, so also did the economics of the entire project become increasingly dubious. But in 1873 the prime minister of Canada, Sir John A. Macdonald, did agree to call for tenders for some parts of works that would have got the canal project started. On 5 November of that year, however, the Macdonald Government resigned and so no work was started. There now appeared the greatest enthusiast of all, an engineer of Woodstock, N.B., named Henry G.C. Ketchum. He was persuaded that a ship's railway would be more satisfactory and economical than a canal and so proceeded to design one. It was to be 17 miles long, almost without gradient, and equipped with widely-spaced double rail tracks, a lifting mechanism capable of raising thousand-ton ships being part of each terminal arrangement. A masonry dock was to be constructed at the Bay of Fundy end, capable of holding 6 vessels. The Chignecto Marine Transport Company was incorporated in 1882, to receive an annual subsidy amounting to $170,602 payable for 20 years. A contract for construction was awarded and orders placed for rails and equipment. Sir Benjamin Baker, another noted British civil engineer, was associated with the work. Progress had to be halted in July 1891 when three quarters of the project was complete: twelve miles of track had been laid; cradles and hauling locomotives were almost ready for delivery. But more funds could not be raised because of the economic crisis of the time, another change in the Government of Canada and, in all probability, the potential that main-line railways were already presenting.[4]

All work stopped. Equipment was disposed of elsewhere, the heavy rails being used on the Intercolonial Railway (from Halifax to Montreal) which would have had to cross the Chignecto Marine Railway. Traces of the excavation may still be seen today but that is all. Ketchum died at the age of 56 in 1896 and was buried, at his own request, in Tidnish close to his beloved railway. The dream remained, however, and has periodically been revived in this century. A delegation of 60 Maritimers was received for further discussion of the project by the prime minister of Canada and eleven members of his Cabinet as recently as 1950. Current estimates of cost approach $100 million so that it is not surprising that careful economic studies all point to very questionable financial implications. The growth of Halifax as one of the great container ports of the North Atlantic coast (as is also Saint John) would seem to sound the death knell for the Chignecto Canal scheme. To stand there on the salt marshes, however, with the sea at both ends of the route almost within sight at once, makes one regret the infallibility of economic studies and explains the hopes which still persist that one day it will somehow be built.

The St Peter's Canal

The Bras d'Or Lakes constitute a unique feature of the physiography of Cape Breton Island, occupying a large portion of the entire area of the south-eastern part of the Island.

The natural entrance is through the magnificent Great Bras d'Or channel on the northeast coast, a few miles to the west of the important industrial city of Sydney. Through this channel 10,000 ton vessels sail regularly to load with gypsum from adjacent mines. The lakes add great beauty to Cape Breton. Being connected with the sea, they exhibit a small tidal range but the narrow entrance naturally minimises this. Never far from the sea, the shore line of the lakes comes very close to it in a bay at the south-eastern end of the main Bras d'Or lake . Here a narrow isthmus on which stands the town of St Peter's separates the lake from a well protected bay of the sea, also known as St Peter's.

So well located is this bay that it was used as a fishing harbour by the Portuguese even before the French regime, being known as San Pedro between 1521 and 1537. From 1650 to 1713, the little port was known as St Pierre but in 1713, at the time of the signing of the Treaty of Utrecht, it was renamed Fort Toulouse. It was attacked and destroyed in the fighting that centred around the great fortress of Louisburg, to the north-east, but rebuilt by the French in 1748. After the second and final fall of Louisburg in 1758, Cape Breton came under the British flag. British settlers came to the small settlement which was again renamed, this time as St Peter's. St Peter's it has remained.[5]

The isthmus is only about half a mile wide. Even though it consists of a fairly high rocky ridge, it was an obvious place for a portage from the sea to the lakes and was so used from the very earliest days of French settlement. Equally obvious was the possibility of cutting a canal through the ridge, once canals had been accepted as channels of commerce. Agitation for such a canal became strong once Cape Breton had joined with the mainland in 1820 to form the united province of Nova Scotia. A saving of from 30 to 35 miles in sailing distance between all points to the south and west and the port of Sydney would be achieved by the canal, the un-favourable winds and currents always found around Scatari Island being thus avoided, an advantage of special consequence in winter. Francis Hall was again engaged to make a survey, his report of 1825 suggesting a total cost of £ 17,150 for a canal 2,700 ft long, 21 ft wide at the bottom and 13 ft deep. An Act was passed by the provincial legislature in 1840 for the establishment of the St Peter's Canal Company but, almost as usual, nothing happened.

In 1850 C.W. Fairbanks made a new survey and even offered to build the canal for £12,500 but his offer was not accepted. In 1853 yet another survey was made, this time by Captain P.J.S. Barry RE, his estimate of cost being similar to Francis Hall's. Work started on 7 September 1854, not by a private company but rather under three governmental Commissioners. Operations were suspended in 1856, however, probably because of rising costs, estimates made in these years rising to £52,000. When completed the canal actually cost £75,000. Work started again in July 1864 but clearly had not advanced very far by 1 July 1867 when all such public works were taken over by the new Government of Canada. The canal was finished in 1869 and came into use immediately.[6]

So useful did it prove to be that it was enlarged between 1875 and 1881 to give a depth of 18 ft. A major reconstruction was carried out between 1912 and 1917 when the single lock was rebuilt to be 300 ft long, 47 ft 4⅞ in wide, with a normal draft over the sill of 18 ft, the canal excavation being 55 ft wide at water line. This is the lock still in use today.[7] It is unusual in that it has double pairs of mitre gates at each end, considering the seven ft tidal range in St Peter's Bay and the negligible range in the Bras D'Or Lakes. It is therefore possible for there to be a difference in water level both up and down when traversing the canal from the seaward end. A main road (to Sydney) crosses the canal at the lake end by means of a simple swing bridge, operated on request by the lock staff with a geared capstan arrangement, the arms for which have to be inserted in the roadway after the bridge has been closed to road traffic.

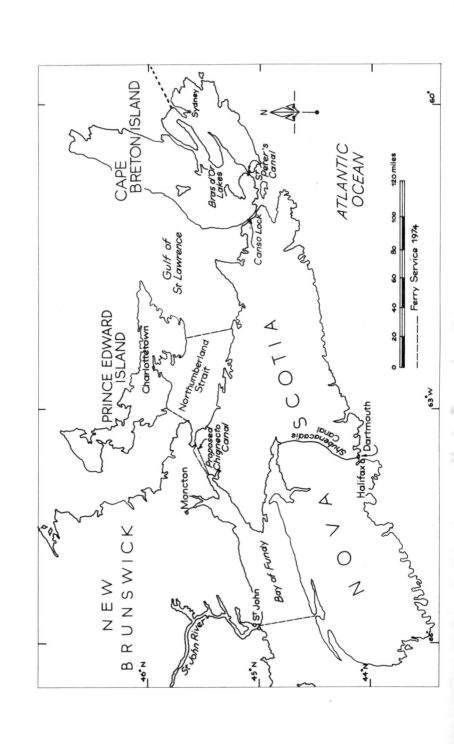

An overall picture of the traffic that the St Peter's Canal has carried down the years is given in Appendix 1. For many years it was used by vessels carrying coal from Sydney mines. The most recent industrial cargo to pass through the canal was a load of oil in a small German tanker en route to North Sydney. It is popularly understood to have proved useful for naval traffic during the years of the second world war but official naval records do not support this understanding. But they do support the local tradition, for such it is, that a Canadian destroyer once passed through the canal into the safety of the lakes. This was done, so I was assured when standing one day by the canal, by waiting until water levels in sea and lake were equal and then opening all the gates when the vessel was passed through, some difficulty being encountered at the bend in the alignment of the canal just after the lock, which is at the seaward end, had been passed.

Through the Director of Information of the Department of National Defence, Canada, I have been kindly informed that the records of the Historical Branch show that HMCS *Patriot* passed through the canal to the sea on 27 September 1925 after gunnery exercises in the Bras d'Or Lakes. Gunnery targets were left in the care of F.W. Baldwin, then carrying out some of his notable aeronautical research work at Baddeck. Exactly four years later HMCS *Champlain* also used the canal, this time for access to the lakes from the sea. A strong wind was blowing and, since the full dredged depth was available only at the centre of the canal, the commanding officer had his ship warped through. The *Champlain* and the *Patriot* were 266 ft 9 in and 271 ft long respectively so that they might possibly have been locked through but the closeness of their dimensions to the overall size of the lock seems to support the local tradition about the passage of a destroyer through the lock with all its gates open.[8]

It is a pleasant story and its persistence locally shows the pride of St Peter's in this small canal despite recent serious reduction in its traffic. It is still used occasionally by fishing boats and, in the months of summer, by many pleasure craft

including some large US vessels bringing visitors to some of
the fine summer homes on the lakes. The grounds around the
canal walls and the lock are beautifully maintained. A small
museum has been opened on the height above the canal
named after Nicholas Denys, a Cape Breton pioneer. The
grounds on the northern side have now been developed into
a most attractive provincial park by the government of Nova
Scotia, with 43 camping sites available for overnight stays.
Isolated though it is, and but little used as compared with
earlier days, the St Peter's Canal remains as a delightful
example of a well planned, if miniature, waterway well maintained and now well served to give pleasure to all visitors
who come to see it whether by land or by water.

The Shubenacadie Canal

Halifax, capital of the province of Nova Scotia, is located
on the south-west side of the narrow entrance channel to
Bedford Basin, one of the finest harbours on the North
Atlantic coast. Facing it, on the north-east shore, is the
smaller but growing city of Dartmouth. The ground rises
rather steeply from the water's edge at Dartmouth; at the top
of the resulting escarpment there are two lakes which form
an attractive feature of the city. They are only two of about
thirty lakes now included within the enlarged boundaries of
the city, one of which comes close to the upper of the two
'Dartmouth Lakes' (the old name; Banook and Micmac are
the names used today). The significance of this detail of local
physiography is that a third lake (originally Shubenacadie
Lake, but now Charles) drains not into Halifax harbour but
to the north, to the head of Cobequid Bay, the upper end of
the Bay of Fundy. The lake is the start of the Shubenacadie
River, which links a chain of small lakes on its way to the
shores of the bay with its great tidal ranges. It is not surprising to find that this natural waterway was regularly used by
the Indians, before the coming of the white man, as a portage
route from the Atlantic shore to the sheltered waters of the
Bay of Fundy. It was a natural consequence that, as soon as

canal fever was experienced in this part of Canada, canalisation of the Shubenacadie route should be proposed.

As early as 1797 legislation was passed authorising the sum of £250 for a survey. This was duly carried out by Isaac Hildrith and Theophilus Chamberlain, who estimated that 19 locks would be necessary. The fervent hopes of the day are reflected in these words of Sir John Wentworth, then lieutenant governor of the province: '[The canal] would be in the interest and convenience of numerous and increasing habitants to purchase of the fair trader in Halifax, whence the frauds, lying, violence and prejudices attendant on illicit commerce will naturally vanish.' An Act was passed authorising the establishment of a canal company but no company was formed. When a new governor, the Earl of Dalhousie, opened the Legislature in 1820 he strongly urged the canal as a desirable public work but it was not until 10 February 1824 that £ 500 was voted for another survey.[9]

Francis Hall comes into the picture again; he was engaged to make the survey and in his report waxed enthusiastic about the practicability of the route. His estimate for a canal with an 8 ft depth was £ 53,344. With a capitalization of £ 60,000 and a grant from the legislature of £15,000, the Shubenacadie Canal Company started work in 1826, the plans calling for 15 locks capable of taking vessels drawing 8 ft. Lord Dalhousie turned the first sod. Once again, Thomas Telford was consulted and was so impressed that, in addition to making a most favourable report (without ever visiting Canada or the site of the works) he appears to have invested £ 450 in the company. All admirers of Telford, and I am but one of many, must be puzzled by his commendation of proposed works the sites of which he had never visited, but he was then almost seventy years of age and must have lowered the standards that he normally followed in his earlier professional career. He died in September 1834 by which time all work on the Shubenacadie Canal had stopped, £ 72,000 having been spent with no more money available. It was never to be completed to the original plans.

The British government had provided some of the money in the form of a loan of £20,000 under certain strict conditions. When these were not met, the canal works were sold under a foreclosure of mortgage and passed into the hands of the provincial government. The British government then agreed to make the loan an outright grant. It is interesting to speculate what the effect of this would have been if done some years previously—the canal might even have been finished as planned. But the provincial government passed the incomplete works to a new company in June 1854, the Inland Navigation Company. A new engineer was engaged who substituted inclined planes for the 5 locks at the Dartmouth end, and for the 3 locks at Portobello, with lifts respectively of 55 and 33 ft, hydraulic machinery providing the power for operation. The canal was apparently finished to these plans and so had 2 inclined planes and 7 locks in its total length of 53½ miles. It had cost $200,000 when opened in 1861, with a small vessel making the round trip to the Minas Basin,but the company was in immediate financial difficulties. The entire project was sold by the sheriff and a new Lake and River Navigation Company formed to operate the canal, which it did from 1862 to 1870.[10]

'As a commercial enterprise, the diminished canal proved a dreadful failure.' This was the judgement of a later city engineer of Halifax whose paper, describing the canal, in the *Transactions* of the American Society of Civil Engineers starts with the words, 'This undertaking, although it has so far proved an utter failure in every respect....' which must surely be unique in the annals of civil engineering. This harsh assessment was, however, correct. Despite the acclaim that greeted the little *Avery* on her initial round trip to Maitland in 1861, traffic did not develop even under the new company, the greatest profit ever made in any one year being $3,000. Some timber and coal was carried on the twelve scows which served the canal, with three steam tugs for hauling, but in 1870 the whole enterprise was sold to L.P. Fairbanks, who continued to operate it for a very short period. He then got involved in a lawsuit about a gold strike on the canal property. While this was in progress two fixed bridges were built across the canal, one for a highway and one for the new railway that had just come to the Halifax area.[11]

That was the end; the canal has not been used since. But some of the locks still remain, their sturdy masonry now mainly hidden in the woods. The waterway is still there also and is used annually by members of some of the active local canoe clubs, two of which have their clubhouses adjacent to the one lock, partially reconstructed, that is seen by the public at the exit from Lake Banook. There are great hopes, locally, that the canal may yet be restored because of its historic interest, and its route developed for recreational purposes. Visitors who have seen its old locks will certainly share these hopes so that, although most Canadians who have ever heard of the Shubenacadie Canal assume that it disappeared long ago, the old canal may yet have a useful but different life ahead of it.

The Canso Lock

Cape Breton is an island, separated from the mainland by a deep, narrow and tide-swept channel, the Strait of Canso. Perhaps one should say 'was an island' since in 1955 the strait was blocked by a large rock-fill dam which provided direct road and rail connections for the first time. Prior to that, ferry services had operated across the narrow strait throughout the year. (The train ferries, necessary after the coming of the railways, were one of the two great fleets of such special vessels that Canada has had to develop and maintain down the years, the other major fleet being that serving Prince Edward Island.) Not only is Cape Breton a lovely place with an important industrial area on its northern coast, but in some ways it is a place apart. The Gaelic is still spoken there; there is even a Gaelic College for the maintenance of Highland culture. Many Cape Bretoners liked their isolation, their feelings well expressed by a Presbyterian minister who is said to have included in his prayers during one Sabbath service, 'We thank Thee, O Lord, for the Gulf of Canso which separateth us from the wickedness that lieth on the other side thereof....' And when the causeway was opened, amid great rejoicing and in the presence of a hundred pipers, one of the welcomes for this link with mainland Canada was the statement that 'We're all Cape Bretoners now'.

Despite such happy sentiments, the train ferries had to be replaced. They carried to the mainland freight cars full of coal and iron and steel products for the markets of North America and supplies for the busy island on their return trips, as well as a heavy passenger traffic including, until recent years, through sleeping cars between Sydney and Montreal. The strait is deep, its tidal currents making ice conditions during winter months hazardous indeed. Since its narrowest section, near to which the train ferries operated, is about three quarters of a mile wide, design of a permanent crossing was an engineering challenge. A bridge could have been built at a great saving in cost over the massive rock-fill dam that was eventually constructed. Opened on 13 August 1955, the dam contains about 10 million cu yd of rock, and cost almost $20 million. 4,500 ft long, it carries a dual lane highway, a footpath and a single track of Canadian National Railways, together with essential services such as telephone and power cables. Shipping had also to be accommodated since the strait was a regular route for vessels from the south coming into the St Lawrence. A ship lock was therefore constructed as an integral part of the causeway constituting, with its short approach channels, one of the most interesting and unusual of the canals of Canada.

The lock is located on the Cape Breton side of the strait, conveniently located for navigation into its approach channels. Construction started in April 1953; it was opened for vessels with drafts up to 10 ft on 1 September 1955 and finally completed by December 1956, commencing full service with the opening of the navigation season in 1957. Length between the ends of approach mooring berths is 0.78 miles, the lock itself being 820 ft long and 80 ft wide providing for vessels with drafts up to 28 ft, although a minimum depth of 32 ft is provided over the roller sills for the two pairs of steel sector gates. Double pairs of gates are again necessary because the tidal difference may be on either end of the lock, depending on tidal movements and winds, the latter sometimes causing a head of 10 ft instead of the normal tidal maximum of 8 ft. All operations are electrical and are interlocked with the operation (by CNR) of the large steel swing bridge which carries railway and road over the southern approach to the lock. Appendix 1 will show that the Canso lock is well used, about three quarters of the vessels locked through being freight or fishing vessels, the remainder pleasure craft. Some of the largest vessels to pass through regularly are those carrying full loads of gypsum from Nova Scotia mines. A pilot service is provided and must be used by all but local vessels, a ship-to-shore radiophone service providing necessary communication.[12]

No reference to the Canso lock would be complete without brief mention of a remarkable but quite unexpected by-product of the building of the causeway. Those who favoured a bridge crossing, of whom I was one, can now be glad that they were wrong, as events have so clearly proved. The depth and narrow nature of the strait have been mentioned. It is just over 14 miles long, its approach from the Atlantic completely sheltered by Canso Bay. Once the causeway was closed, ice flow through the strait in winter months was stopped. Tidal conditions lead to an ice pack on the St Lawrence side of the dam but clear water on the other and hence down the full length of the strait. This is now being developed as one of the greatest deep-water ports in the world, there being already in use one major oil wharf with 100 ft of water alongside only 750 ft from the shore, capable of handling, as it has done already, the largest oil tankers afloat. Rarely if ever has the construction of a transportation facility had such unexpected and beneficial side-effects.[13]

CHAPTER 3

Water Route to the States
The Chambly Canal
and the Lock at St Ours

History and geography combine again in a somewhat surprising way to explain the existence of the Chambly Canal in Canada and the associated Champlain Canal of the New York State Barge Canal system. Lake Champlain may be seen from a map to be a long narrow lake, oriented almost due north and south, lying between the Adirondack Mountains of New York State and the Green Mountains of New England. About 100 miles long with an area of 600 square miles, its size always surprises visitors but readily explains the naval battles fought on its waters. It drains to the north through Canada, down the Richelieu River which joins the St Lawrence at Sorel, the river flowing also due north. To the south of the lake, its valley continues, soon to be the valley of the Hudson River flowing almost due south to enter the Atlantic Ocean at New York. From Whitehall at the foot of Lake Champlain to Fort Edward on the upper Hudson River is only 23 miles. Nature has thus provided one of the great water routes of North America, of special significance in that it links Canada with the United States and, in particular today, the two great metropolitan cities of Montreal and New York.

The Indians knew this route well and so introduced it to the earliest white settlers. Champlain, the great and good man who was the real founder of Canada, accompanied an Indian expedition up the Richelieu River in 1609, only six years after the establishment of his first settlement at

Quebec. They reached the lake that so rightly bears his name on 14 July of that year. The lake was the scene of many Indian fights, in one of the most serious of which Champlain was wounded. This was in 1615; the great French soldier-explorer got back safely to his *habitation* in Quebec in 1616 by way of the route followed by the Trent Canal (p 86) and the Ottawa River (p 118). But as early as 1666 the Richelieu River was being used for peaceful transportation, large scows regularly sailing between Sorel and Chambly with only one portage, this being at St Ours where rapids impeded direct sailing up the river. Between Chambly and Fort St John's (St Jean, today) there were 12 miles of almost continuous small rapids, necessitating the second and only other portage on the river between the St Lawrence and Lake Champlain. French military authorities constructed rough portage roads around these two impediments to navigation. In 1747 they also removed the 'chain of rocks' which used to distinguish the rapids at St Ours, this being the first dredging operation known to have been carried out in Canada.[1]

The Champlain route was therefore an obvious one for canalisation. The first proposal appears to have been made by Silas Deane of Connecticut in a submission to General Haldimand, governor of Quebec; this was repeated in 1787 to a new governor, Lord Dorchester, but still with no result. In 1792 Colonel Schuyler's Company was incorporated in the United States for the construction of the Erie Canal and the Champlain Canal, the latter to connect the Hudson River with Lake Champlain. This naturally reactivated interest in Canada, Ira Allen being particularly busy in attempting to interest British authorities in constructing the essential canals on the Richelieu River if the through route was to be achieved. His efforts met with no success. It was not until 1 April 1818, following the War of 1812 (about which we shall have more to say in the next chapter) that the Legislature of Lower Canada passed an Act incorporating the Company of Proprietors of the Chambly Canal. Locks were to be 20 ft wide, providing a 5 ft depth of water.

Supervising Commissioners were appointed. Surveys were made but construction did not start until 1833. By 1835 work was entirely suspended. It was not until 17 November 1843 that the little canal was completed, work having been resumed after the union of Upper and Lower Canada in 1841 under the newly constituted Board of Public Works, an official agency of the new government. The Board also constructed the necessary regulating dam and lock at St Ours in 1849 but gave this lock dimensions of 200 ft long by 45 ft wide with a depth over sills of 6½ ft, clear indication of the advance in thinking about canal traffic by the 1840s.[2]

The Champlain Canal in the United States had been completed in 1825. It will naturally be described in the companion volume in this series, *The Canals of the United States*, but since it is so closely associated with the Chambly and St Ours canals, it must be briefly noted here. It now consists of 23 miles of artificial waterway and 37 miles of an improvement of the upper Hudson River, having 11 locks, a rise of 43.5 ft from Lake Champlain to its summit which is fed by the Glen Falls Feeder Canal, and then a descent of 138.5 ft to its junction with the Hudson River. Naturally reconstructed since its original building, its locks have provided a depth of 12 ft over sills, locks being 45 ft wide and 328 ft long, throughout the present century. It continues to serve the city of Burlington and other municipalities on Lake Champlain, but traffic through its locks naturally consists now mainly of pleasure craft. The dimensions of its locks are significant, as will soon be seen.[3]

The Chambly Canal is 11.78 miles long, having 9 locks capable of giving a total lift of 80 ft. Locks are still 120 ft 7 in long and only 23 ft 7 in wide, the depth over sills being nominally only 6 ft 6 in. The actual clearance depends on the level of the Richelieu River; it has been as low as 6 ft at the lock at the downstream (Chambly) end. Low water levels in the River have sometimes limited the draft of vessels making use of the Richelieu route so that a good deal of dredging of the river bed has been done, a major programme having been carried out as recently as 1935. Despite the limited size of the locks, the canal carried heavy traffic during the middle and latter part of the nineteenth century, especially of barges carrying sawn lumber from the forests of Canada to markets in the United States. Many of these barges came down the Ottawa River (see p.68) to Montreal and then down the St Lawrence to Sorel where they turned into the Richelieu for their journey south. Before the days of railways, this was the route also followed by passenger vessels. The short-cut, so obvious from the map, from St Jean to La Prairie opposite Montreal, was naturally put to use at an early stage of modern travel, St Jean becoming the important transfer point from river steamer to stage coach. It is not surprising to find that this 'portage' was also the route of Canada's first railway, the Champlain & St Lawrence Railroad, opened in July 1836, but still a supplement to the steamer service from St Jean, up the Richelieu River and through Lake Champlain. This was the route followed by Charles Dickens after his visit to Montreal in 1842 during his first North American journey, so interestingly but briefly described in his *American Notes*.[4]

It was inevitable that proposals for a new canal along the same route as the railway to Montreal should have been suggested, to connect with the Chambly Canal and so give direct access from Montreal to Lake Champlain without the long detour by way of Sorel. The saving in distance between Montreal and St Jean would have been about 70 miles. The initial railroad naturally had very limited capacity and traffic was increasing on the Chambly Canal so that the advantages of the additional water connection to Montreal were 'obvious'. No less than five different schemes were suggested, one being for a canal across the relatively level plain of the St Lawrence valley from the Beauharnois Canal (see p 154) to the Richelieu at St Jean. The discussions came to nothing and so the lock at St Ours and the smaller locks of the Chambly Canal continued their good service down through the century.

The suggestions for improvement and enlargement, however, never stopped. It is fair to say, even though it could not actually be proved, that probably no other canal has been the subject of so many high-powered political discussions, proportional to its size, as that at Chambly. Yet it remains in service with its original masonry locks, found during remedial work to have been founded on timber grillages, still in reasonably good shape as a result of sound building and good maintenance. Little would be gained by any detailed recital of the political discussions about the future of the canal but their character may be judged from the fact that as recently as the early 1960s yet another proposal for a canal link to La Prairie was quite seriously proposed.[5]

The continued political pressures and public discussions did, however, achieve one definite result when the original St Ours lock and dam were replaced by new and more modern structures, completed in 1932. The early thirties were tragic days for Canada, as for all of North America, in view of the economic depression coupled with the absence, then, of the social measures that today would ameliorate to a large degree the human distress of those years. Public works were used to provide some work for the vast numbers of unemployed. It appears to have been in this way that the decision was made to rebuild the old structures at St Ours. The fact that the works were initiated by a Minister of Public Works who happened to be the Member of Parliament for Sorel may possibly also have had some connection with the work. An entirely new concrete dam with necessary steel control gates was built in the west channel, there being a small island that here divides the river into two, and a splendid new ship lock was constructed in the east channel. The new lock is 339 ft long and 45 ft wide; it has a depth of water of 12 ft over the sills, thus corresponding with the ruling dimensions of the Champlain Canal at the US end of Lake Champlain. A 12 ft channel was also dredged up the river as far as Chambly Basin, the entrance to the Chambly Canal, but there the improvement stopped.[6]

The international character of the Richelieu River, draining as it does an important lake in the United States, has long given this 'Champlain Route' (as the entire waterway is usually designated) a special international significance. Waters that cross the boundary between Canada and the United States being under the aegis of the International Joint Commission (IJC), supervision of the Richelieu River and Lake Champlain is one of the Commission's continuing responsibilities. In 1936 IJC was given the task of studying, once again, the possibility of establishing a true deep waterway from Montreal to New York, using the Champlain Route. Public interest was so aroused that a delegation representing no less than 150 municipalities in the Richelieu River area waited upon the Canadian Minister of Transport when he visited St Jean in February 1937. Equally vociferous were the opponents of the scheme, some politically motivated, some representing the great railroad interests which were quite bitterly opposed to any improvement of waterways such as the Chambly Canal (as well as the much greater St Lawrence canal system, as we shall see in Part Two of this volume).

Eventually IJC decided against any major improvement of the Chambly Canal but did authorize the construction of a cross-river dam fitted with steel control gates at Fryer's Island, six miles downstream of St Jean. This was completed just before the outbreak of World War II, which naturally diverted attention away from such domestic works. As a result the companion control dykes along both banks of the river upstream of the control dam, without which it could not be placed in operation, were not constructed, nor have they been built to this day, so has interest in the Champlain Route waned. The dam would have given some control over the level of Lake Champlain, especially at the time of spring floods, and with associated dredging would have given deeper water for navigation in the 22 miles between St Jean and the international boundary, i.e. the exit from Lake Champlain. The dam is in good shape, the steel control gates being regularly operated as a part of annual maintenance procedures, so that it is available for use should the decision ever be made to improve this interesting and historic water route.[7]

Despite the pressures of Canada's war effort, time was taken by members of the Government of Canada to attend the impressive ceremony held on 5 September 1943 at the foot of the canal in Chambly Basin to mark the centenary of its opening. Naturally, the opportunity was not lost for putting forward strong pleas for improvement of the waterway, at the very least for deepening the Chambly Canal to give a depth over sills uniform with the Champlain Canal at the other end of the lake. It was pointed out, not for the first time, that the distance from Montreal to New York by the Champlain route was only 452 miles as compared with 1,670 miles by the usual sea route down the Gulf of St Lawrence and along the Atlantic coast. All to no avail; polite replies were made but no commitments given and so nothing happened, again. Fortunately, proposals for the abandonment of the canal, quite seriously urged by its opponents, have also not been heeded. Maintenance has always been well done and so the canal now serves the many pleasure craft that use it to such good effect in summer months.[8]

For a period of just over twenty years an enterprising Canadian company showed what the canal could do for commerce, despite its small size. Until 1936 the wooden scows then used for such freight traffic as the Champlain route did carry were hauled by horses through the Chambly Canal, the only regular use of horses for canal work in Canada during this century. Seven small steel diesel-powered vessels were then constructed, specially designed so that they would just fit in the small Chambly locks with a minimum loss of capacity. They started operating a regular freight service in 1937. Each vessel was capable of carrying 230 tons, their average load being always well over 200 tons. They loaded newsprint at Donnaconna on the St Lawrence for the use of New York publishing companies, bringing full

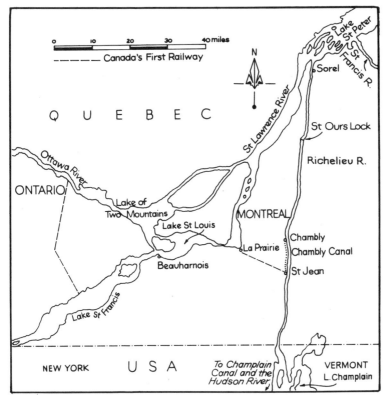

Map B: Water Route to the USA

cargoes of coal, raw materials and package freight on their
return north-bound voyages. A one-way trip took 2½ days.
So good was this service that in twenty seasons of operation
well over half a million tons of newsprint valued at about $50
million was safely delivered to New York.[9] A number of
factors combined eventually to lead to the cessation of this
little service which finally terminated in 1959. The fact that it
was in operation for so long as a private venture (despite the
inevitably high ton-mile cost with such tiny vessels) gave
some encouragement to those who still hoped to see the
Champlain Route improved. More cold water, however,
was thrown over all such hopes by the report of one of

Canada's innumerable Royal Commissions, this one known as the Glassco Commission from the name of its chairman. Reporting in 1963, the Commission suggested that tolls be charged again on Canadian canals and that as 'expensive anachronisms' they be passed over to provincial governments. Most fortunately, this particular recommendation was treated with the complete neglect which it deserved; the commissioners must all have been devotees of the automobile and assuredly not canal lovers.[10]

Let those of us who are canal lovers take an imaginary journey down the Canadian part of this historic Champlain route. This can start at Lake Champlain with a pleasant easy sail, if the wind is in the right quarter or our auxiliary engine working well, down 22 miles of broad river, well buoyed. Quiet and rustic as are now the bank-side scenes, this same stretch of water was the passageway for many an armed force as we shall be reminded when passing Ile-aux-Noix, one of the strategic points on the Richelieu. Buoys will guide us to the left bank as we approach the busy industrial centre of St Jean, a large marina being located adjacent to the entrance wharf to the canal where necessary marine supplies may readily be obtained. Close study of the two banks will reveal remains of a railway line long operated by Canadian National Railways, that once crossed the river here by a bridge with a swing span over the canal. It is all now a thing of the past, the bridge having been dismantled as a part of railroad rationalization. After noticing a reconstructed guard gate we shall pass under, or through if clearance is insufficient, two bridges—one carrying a main road and one a main line of the Canadian Pacific Railway, both with swing spans. The first lock is located just downstream of the CPR bridge. Its small size will at first surprise the visitor, especially if he remembers the fleet of diesel freighters that used to operate through it, but surpise will turn to pleasure as the well kept grounds and gardens around the lock are enjoyed.

For the first 10 of its 12 mile length, once the first lock has been passed, the canal is a level stretch of water, 75 of the 80 ft total drop taking place in the last 1½ miles at the Chambly end, through careful location. The canal is separated from the now swift-flowing river by a broad embankment, Quebec Highway No. 47 running parallel on the high bank above the canal. After about three miles, the canal begins to depart from the river proper, at first using what was the west channel around Ile Ste Therese but then located in an artificial cut which remains on the higher ground at an increasing distance from the river, the canal being crossed by the highway about 8 miles from St. Jean. Thereafter the canal is lost to view from the main road but can be reached (if we switch now to an inspection on land) by taking one of the small farm roads to the west, some of which are provided with simple swing bridges for crossing the canal. When the canal is reached by going up one of these side roads, it can again be followed, certainly on foot or in a small car, all the way to Chambly, by using the broad old towpath along which horses used to haul barges until forty years ago. It is in this secluded stretch of the canal that the next five locks are located, the lower three being relatively close together but with basins between each. A pleasantly designed small lockmaster's building is located at each of the locks, all of which are maintained immaculately as are the surrounding gardens. With no road traffic in view, this pleasant sylvan section of the Chambly Canal is the only place in Canada that resembles the accustomed European canal scene, especially so on a lovely day in high summer when, with a blue sky overhead and the flat verdant country around, one can readily imagine oneself on a canal in the Low Countries.

A modern railway bridge was passed just after the first of this group of 5 locks; this carries a main line of Canadian National Railways. At the foot of the fifth lock, the canal is approaching the town of Chambly, swinging around before being crossed by another main highway bridge located at the top of the flight of three locks that finally takes the canal into the river again in the south-west corner of a wide expanse

known as Chambly Basin. The highway is a busy one, leading to a major road crossing of the Richelieu River, and the
town is a virile developing modern community. The activity
around and near the flight of locks is, therefore, in great
contrast to the quiet and rustic character of the canal only a
mile or so upstream but this adds to the undoubted charm
which the Chambly Canal exercises upon most of those who
know it. Another large marina adjoins the foot of the flight of
locks, again with necessary marine services readily available. A stop here is desirable, if time permits, since only a
short walk away from the locks is the reconstructed Fort
Chambly. It occupies an obviously strategic position at the
foot of this long stretch of rapids on the river. Built originally
as a log fortress in 1665 as a defence against the Iroquois
Indians, who used the river as their route for attacks on the
early French settlements, it was rebuilt as a stone fort in
1709. This was as protection against the advance of British
armed forces coming up from Albany, again using the river
as their route. Ruins of this structure are the basis of the
present reconstruction, the stone fort having been next used
by the British as a defence against the US troops in the War
of 1812 and in 1837 as a prison for some who took part in the
insurrection of that year. It is indeed a place of history.

   The fort and its pleasant grounds will be passed if we
continue on the most convenient road for our land inspection
of the Canadian part of the Champlain Route. This involves
crossing to the right bank of the river by the modern bridge
just beyond the fort, then making a sharp left turn onto
Highway 21 which leads all the way to Sorel. The road runs
close to the river, separated usually by almost continuous
well developed river-front properties, the river bank providing quite delightful residential settings. The river itself is a
broad smooth-flowing stream of singular uniformity all the
way to St Ours, a distance of over 30 miles, its level
controlled by the dam that has there been built in the western
channel. A 12 ft deep channel has been dredged throughout
this full 30 mile stretch, which is well buoyed, but only the

occasional *shalop* or small trading vessel from the St Lawrence is to be seen moored at one or other of the small wharves located along the river. One comes to the lock at St Ours quite suddenly, because of turns in the road. It is close to the right bank and so is readily approached from the road we have been using. Its size strikes one immediately, in such contrast to the smaller Chambly locks, reminding one vividly of the hopes to have such larger locking facilities all the way from the St Lawrence to the Hudson. Again, and one may correctly say 'as always', the lock, its equipment and the grounds around are maintained in beautiful condition. One may cross one of the gates and so gain access to the island, gaining a good view here of the control dam in the other channel. And a pleasant way to finish a visit to this waterway is to use the simple ferry scow just downstream of the lock to cross back to the left bank, this providing a short cut to the main road up the St Lawrence to Montreal.

The combination of the fine St Ours lock and the venerable Chambly Canal, with its much smaller locks, provides one of the real paradoxes of the canals of Canada, more particularly so when it is remembered that they connect with the Champlain Canal in the United States which is also provided with locks of the larger dimensions. If, as some think, inland water transport in Canada is a thing of the past then these public works on the Richelieu River will probably remain as they are today, well maintained at relatively low cost, merely serving a steadily increasing volume of pleasure boat traffic. But if, possibly in view of the world-wide concerns about energy that have characterised the mid-seventies, and in view of the phenomenal increase in inland waterway freight haulage in the United States, the Champlain Route between Canada and the United States is ever seen to be economically viable and internationally desirable, it will be a relatively straightforward engineering task to convert the Chambly Canal so as to give the standard 12 ft depth of water throughout the route. In view of the energy

crisis that is now recognised throughout the world, rehabilitation of the Chambly Canal would appear to warrant serious consideration. The hopes of a century and more therefore remain; they may yet prove to be hopes fulfilled.

CHAPTER 4

Canals for Defence
The Ottawa River Canals
and the Rideau Canal

The War of 1812, often called 'Mr. Madison's War', was one of the most lamentable conflicts of recent times, a war fought in British North America that had profound influence upon canal building in Canada. It should never have been started, and probably would not have begun if modern trans-Atlantic communication had been available, since the British Government had decided to remove the main irritants of the United States some days before President Madison declared war. The last unfortunate battle, that of New Orleans, was fought after the Treaty of Ghent had been signed and peace restored, again due to lack of communicaton. Canadians who have studied the war are generally agreed that it was the British forces, then defending the 'colony', which on balance were the more successful. Citizens of the United States are equally sure that their side won—good indication that the result of all the border fighting was really a stalemate. The United Provinces of Upper and Lower Canada were invaded; so also were the United States. Naval battles were waged on Lake Champlain and on Lake Ontario, two US armed schooners sailing down the Richelieu River as far as Ile-aux-Noix (p 44) where they were captured and immediately converted into British naval vessels. Later a frigate was built at Ile-aux-Noix and was the Royal Navy's principal vessel in the ill-fated battle of Plattsburg. Sir James Yeo on Lake Ontario built a three decked ship at Kingston that was more powerful than Nelson's flagship at Trafalgar.

Kingston, at the east end of Lake Ontario, was its naval base and vital fortress.

All supplies for this strategically located fortress had to be brought up the St Lawrence to Montreal and there transferred to river *bateaux* for the long and arduous journey to the safe harbour of Kingston up the St Lawrence River with its many rapids. A glance at the map will show that the latter part of this journey was along the international section of the great river, suggesting the possibility of ambush by US forces. No ambush ever did take place during the war, even though such complex shipments as some prefabricated small naval vessels were brought up this difficult route. The reason was not difficult to appreciate since, just as the Canadian banks of the St Lawrence were untouched forests with no access to them except by the river, so also was the US side of the international section of the St Lawrence merely wild forest with no road links with the south. When the war ended, there remained such strong feelings on both sides that British authorities, not to mention those of the United States, considered that they must be ready for another outbreak of hostilities. Discovery of a US plan for an ambush attack on any further military shipments up the St Lawrence to Kingston merely served to confirm the necessity of an alternative route between Montreal and Kingston as a matter of military and naval urgency. The Duke of Wellington was particularly concerned that such an alternative route be provided, especially after he became Master General of the Ordnance. The Rideau River provided the main part of an essential link between the Ottawa River and Lake Ontario. It was known to the Indians, who were able to show early surveyors alternative connections between the Rideau and Kingston. But the Ottawa, the Rideau and the Cataraqui would have to be canalised in order to be useful; so started some of the most interesting early canal building in Canada, the main result of which—the Rideau Canal—is still in active use.

The Ottawa River is the main tributary of the St Lawrence, a magnificent stream with an average flow greater than that of all the rivers of England and Wales combined. In its course of 700 miles it drops a total of 1,100 ft of which 400 ft occur in the last east-flowing section of the great river, that most familiar to travellers. Most of this 400 ft fall used to be in the great rapids that characterised the river west of the present city of Ottawa where the Rideau River joins the Ottawa. Between this point and the St Lawrence, therefore, there is a drop of between 60 and 70 ft (depending on the state of river flow) and all but a few feet of this was taken up in a 12 mile stretch of rapid water, the main part of which was long known as the Long Sault. Canalisation of the Ottawa, therefore, consisted of the construction of a lock at its junction with the St Lawrence, a place known (rightly) as Ste Anne de Bellevue, and a series of small canals to circumvent the Long Sault and the adjacent rapids, initially a series of three small canals known collectively as the Ottawa River Canals. From Ottawa to Kingston by way of the Rideau River, the Rideau Lakes and the Cataraqui River is a distance of 123.5 miles. Canalisation of this route was a major undertaking but it was successfully completed in little more than five years, between 1826 and 1832. This summary statement will indicate the superb example of civil engineering that the building of the Rideau Canal proved to be, one of the great legacies of the Corps of Royal Engineers to Canada.

The Ste Anne's Lock

The Ottawa River joins the St Lawrence through four channels, as may be seen from the accompanying sketch map. It will there be noted that Montreal is on an island, of the same name as the city, an island formed by the Back River (Rivière des Prairies), with a second smaller island formed by the Rivière des Mille Iles, now the location of the rapidly growing second city of Laval. At the west end of the island of Montreal, the Ottawa discharges into the St Lawrence through two channels, separated by Ile Perrot. The combined drop of the rapid at Ste Anne's and the Lachine

Rapids was taken up in the two island-forming rivers in a series of rapids, now generally flooded by the impounding of water by a water power plant on the Back River and by a control dam on the other outlet. Flow through the power dam now accounts for about one half of the total flow of the Ottawa but before construction of the plant, the rapids at Ile Perrot took most of the total flow. With a drop of up to 5 ft (at high water), these rapids therefore constituted a real impediment to navigation.

Until the early part of the nineteenth century, canoes and other vessels had to be hauled up the rapids with ropes but craft going downstream would usually be guided through the

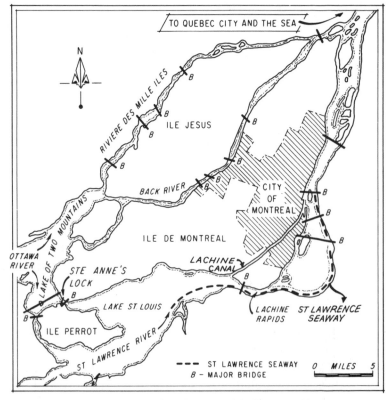

Map C: Junction of the Ottawa and St Lawrence Rivers

rapids by their expert crews, as also those going down the much larger and more dangerous Lachine Rapids. When settlement started on the Ottawa, still in a small way in the first two decades of the new century, regular boat traffic was started between Lachine and the Lake of Two Mountains, first with *bateaux* and then by early steam boats. Some improvement was urgently needed at the rapids and this was first provided by a simple wooden lock at Vaudreuil, adjacent to the mainland on the south. This was built as a private venture in 1816 by the St Andrews Forwarding Company. It provided for small vessels with a maximum draft of 5 ft.

The owners tried to maintain this lock as their own private preserve but their monopolistic position was challenged when a venturesome river man discovered a safe channel up the adjoining rapids and then managed to haul up a fully laden barge by 'snagging' against good anchors. The wooden lock was therefore made available for general use but its inadequacy soon became evident. Many were the studies made for its replacement. The officers commanding the construction of the Ottawa River and Rideau Canals were brought down from their respective stations as members of a committee of inquiry but even their reports (one of which suggested that locks should be built on the Back River instead of at Ste Anne) did not result in any action. It was not until the time of the union of the two provinces (1841) that work did commence on the construction of an entirely new lock, of masonry, and at Ste Anne's. Completed by the Board of Works of the United Province of Canada in 1843, as a public work instead of a private venture, the original lock at Ste. Anne's was 190 ft long and 45 ft wide with provision for vessels drawing 6 ft. This lock was well used, the old wooden structure at Vaudreuil naturally falling into disuse, no trace of it remaining today. But as canal fever mounted, and the clear need for rebuilding the Ottawa River Canals was officially recognized, reconstruction of the Ste Anne's Lock was decided upon and carried out between 1878 and 1886. A parallel and adjacent site was utilised, the new lock being 200 ft long by 45 ft wide with a normal draft over sills of 9 ft. This has varied, between periods of high and low water on the Ottawa, from as much as 20 ft on the upper sill to almost 8 ft on the lower sill.

This is the lock that is still in use today. It has been well maintained and is now operated and lighted by electricity. The normal lift it provides is only 3 ft but, again, at periods of high and lower river flow, this may vary up to over 5 ft. Well built approach walls facilitate passage through the lock, the total length of the resulting little 'canal' (as it is officially described) being 0.12 miles. It is crossed near its downstream end by a group of bridges carrying CNR, CPR and two parts of a modern highway, so the lock is very much in the eye of the travelling public. Good clearance is provided beneath the bridges, normally 41 ft 5 in but varying with river stages between 33 ft 1 in and 42 ft 7 in. The lock naturally provided passage for all vessels proceeding up the Ottawa River, either up to Ottawa or beyond this into the Rideau Canal, in each case after passage through the Ottawa River Canals.

The Ste Anne Lock carried in addition such traffic as originated or terminated at the small wharves around the lovely Lake of Two Mountains. In earlier years this was generally package freight but in more recent years it consisted mainly of large barges loaded with the excellent sand that was obtained from lakeside pits, going down to Montreal to provide some of the aggregate for the mounting quantities of concrete used in the building of Canada's great metropolis. Through traffic going upstream enjoys a clear 27 mile sail across the usually smooth waters of the Lake of Two Mountains as far as Carillon where a great power dam now controls the river. Passage past the dam is made by means of a fine new ship lock, electrically operated. Construction of the Carillon Dam has flooded out, almost completely, the three Ottawa River Canals to which our attention may now be directed.[1]

Page 55. *Upper Canada Village's replica of an early wooden lock built by the Royal Engineers (of Great Britain) on the first St Lawrence canals in the closing years of the 18th century*

Page 56. (Above) *A reconstruction of the first lock on the Great Lakes, at Sault Ste Marie, Canada, at the outset from Lake Superior. Surveys for the lock were made by the North West Company in 1797; it was probably in use at the turn of the 18th century; (below) 11 September 1965: the last arrival at Bagotville of the S S Tadoussac, one of Canada Steamship Line's fleet of passenger steamers that provided a daily summer service between Montreal, Quebec City and the Saguenay River*

Page 57. (Above) *Looking into the eastern entrance of the St Peter's canal, Cape Breton Island, Nova Scotia. This 1886 view from the sea shows the single lock with its double gates; (below) Drawings of machinery proposed for the Chignecto Ship Railway, planned but not completed as an alternative to the long discussed Chignecto canal near the Nova Scotia-New Brunswick boundary*

Page 58. Masonry of Lock No 7 of the Shubenacadie canal, Nova Scotia, as it is today after more than a century's abandonment

Page 59. Canso canal and lock as seen from the air, looking eastwards along what used to be the Strait of Canso but is now an enclosed harbour.
Foreground: top of the embankment that crosses the strait with a road and a Canadian National Railways line

Page 60. (Above) *MV Ethel Tombs approaching the downstream entrance to the Chambly canal in Chambly Basin, on its way to New York southwards with a load of newsprint. This was one of the small fleet of vessels specially designed to fit into the small locks of the Chambly canal; (below) Chambly canal, Quebec; Lock No. 8 looking downstream. A bridge carries Canadian National Railways across the canal in the background.*

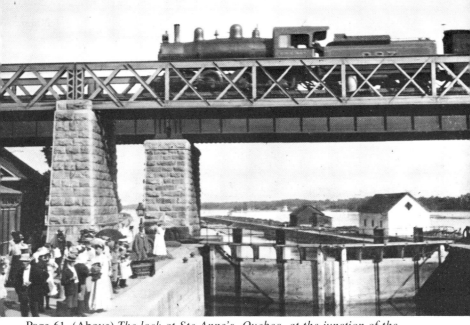

Page 61. (Above) *The lock at Ste Anne's, Quebec, at the junction of the Ottawa River and the St Lawrence in Lake St Louis, photographed about 1900. A train of the Grand Trunk Railway crosses the bridge in the background, eastbound for Montreal; (below) A summer scene on the downstream lock of the double Carillon canal in 1960, shortly before it was replaced by the lift lock constructed with the Carillon Dam and power plant upstream. The masonry is that used in the reconstruction of the canal in the 1870s*

Page 62. The entrance locks to the Rideau canal from the Ottawa river (foreground), as seen by British artist W.

The Ottawa River Canals

Prior to the building of the great new dam, and before a low timber crib dam had been built just upstream to assist the passage of the rafts of squared timber that were floated down the Ottawa throughout last century, travellers up the river were faced at Carillon by a short stretch of rapid water. There was another small rapid two miles further upstream at a location known as Chute à Blondeau, the river flowing swiftly but smoothly between the two rapids and above the second as far upstream as Grece's Point. This used to be a spectacularly beautiful section of the Ottawa River but early travellers had little time to admire the scenery since the roar of the Long Sault would be heard even before they reached Grece's Point. For the next six miles the river was a series of foaming rapids, their beauty enlivened by several small islands that added to the danger of shooting these rapids by canoe, a practice regularly followed down the years by those expert in river lore. There were from the earliest times of river travel rough portage roads along both banks of the river, circumventing the 12 miles of rapids. Those usually used, as well as the original Indian trail, in all probability were on the north bank. Conveyances used the roads on both banks in the early years of the nineteenth century but these merely assisted passengers; the boats they used had to be hauled up the two small rapids and the long 6 mile stretch by man-power. Several pages could be well filled with eye-witness accounts of the hardships and hazards that this passage of the Long Sault involved. It must suffice to say that it was probably the most difficult to pass of all the impediments to navigation on the Ottawa and yet everyone going upstream had to overcome this hazard.

If ever an alternative route to the fortress of Kingston was to be developed, these three sets of rapids would have to be bypassed and this meant the construction of small canals. Military surveys were therefore made of the possibility of building such canals, starting in the first decade of the nineteenth century. The lie of the land pointed to the north

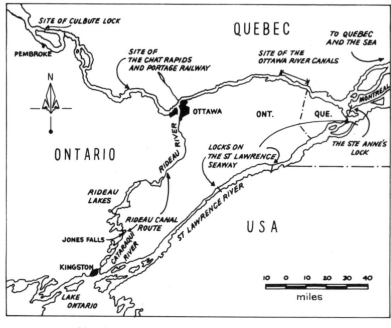

Map D: The Ottawa and Rideau Route to Kingston

shore as the preferable location, the local geology resulting in a flat slope from the water's edge for an appreciable distance back into the woods which came right down to the river bank. Geology provided the corresponding disadvantage of giving solid rock (limestone, generally) almost at the surface of the ground in many places so that it was recognized from the first that canal construction would inevitably involve a good deal of rock excavation. The need for the canals became more urgent after the close of the War of 1812 but, although a further survey was made, the project was not given the highest priority by the British Government. There was the further complication of payment for the canals, when built. The British Government wished to have the governments of Upper and Lower Canada at least share in the cost of construction but the Legislatures of the time were, quite naturally, not disposed to use any of their meagre funds on a project so clearly linked with defence—still the unquestioned responsibility of Great Britain. The impasse was broken on humanitarian grounds, so the records suggest, after the arrival in Canada of the Duke of Richmond who took office as governor on 30 July 1818.

This enlightened nobleman made it his business to travel through the two provinces and to develop a broad interest in the start of settlement. He therefore heard of the hardships and difficulties of a group of over 300 discharged soldiers and their families in getting past the Long Sault on the Ottawa, on their way to establish a new settlement in the forest (now known as Richmond, about 20 miles from Ottawa). He knew of the proposed alternative military route between Montreal and Kingston. Such was his interest that he planned his travel in the summer of 1819 so that he could traverse the route of the Rideau Canal between Kingston and the Ottawa River, even though this meant a journey through wild forest, much of it on foot. He dined with the retired military officers in a simple inn at Richmond near the end of this part of his journey but died tragically the next day (28 August) as a result of a bite he had received earlier on this journey from a tame fox, which must have had rabies. The final part of his planned journey, down the Ottawa River to inspect the proposed Ottawa River canals, he did not therefore make, his body being brought down this same route in a plain deal box for eventual burial in the Anglican Cathedral in Quebec City. But even before his intended visit, he had given orders (anticipating approval from London) for a start on the building of the Grenville Canal which would circumvent the worst part of the river barrier, the Long Sault.

These orders were given to the Royal Staff Corps, a little known regiment of the British Army. Founded in 1799 by the Duke of York, then Commander in Chief of the British Army, to supplement the work of the Corps of Royal Engineers, the Staff Corps lasted only until 1829 when the great economy move of the time led to its abolition. In its short history, however, it served with distinction in the Peninsular

War under the Duke of Wellington; it built the Royal Military Canal, still to be seen between Winchelsea and Hythe in the south of England; and it was responsible for the construction of the Ottawa River canals. This last contribution must be stressed since it is still popularly believed in Canada that the Royal Engineers not only built the Rideau Canal but also the Ottawa River canals. The fine bronze commemorative plaques at each end of the original route of the latter, erected by the Historic Sites and Monuments Board of Canada, even repeat this incorrect information!

Men of the Royal Staff Corps were moved up from work on the defences of Halifax, Nova Scotia, to work on the start of the St Lawrence River canals (see p 153) and from there came up to Grenville for what was to be their main contribution to the opening up of Canada. The Commanding Officer, from the start in 1819 to the completion of the three little canals in 1834, was Lieut-Col Henry du Vernet, a fine man of whom very little is known apart from his patience and devotion in directing the building of the canals. Partly because of the lamentable death of the Duke of Richmond, du Vernet started with a minimum of guidance and no plans or specifications to assist him. His engineering intuition, however, guided him to a good overall plan for the Grenville Canal—a guard lock at the upper end to take care of fluctuations in river level, then a long level stretch of canal proper taking advantage of the lie of the land, concentrating as much as he could of the total drop of 43 ft in locks close to the downstream end. The forest had to be cleared first for a camp site, and then along the route selected for the canal, close to the river bank. These simple operations took up most of the first summer. Work proceeded in succeeding summers rather slowly since du Vernet kept closely to a total expenditure for each year of only £ 8,000, this being the limit set by the Treasury. It is small wonder, then, that the relatively small and simple canal works were far from complete when inspected by a special military commission sent out from England in 1825.

As a result of their recommendations not only was work started on the Rideau Canal, as we shall shortly see, but work on the Grenville Canal was intensified and orders given for the start of work on the two smaller canals at Carillon and Chute à Blondeau. Colonel du Vernet came back from his stay in England to direct all three works. The Chute à Blondeau Canal consisted of a straight rock cutting about 800 ft long with a single lock at mid-point. It was built by contract, as was also the Carillon Canal, although with some difficulty because of divided responsibility between the Staff Corps and the Commissary General and also, perhaps, the unfamiliarity of the Staff Corps with contract work. Colonel du Vernet used his experience on the Grenville Canal to good effect, however, in laying out the line of the Carillon Canal. In order to minimise rock excavation, which was naturally a very slow operation, he followed the lie of the land around the Carillon Rapids, even though this involved locking boats *up* to the main part of the canal on their downstream journey. The route he chose permitted the combined total drop of 23 ft to be taken up in a staircase pair of locks near to Carillon village, leading directly back into the Ottawa River. The drop in each direction necessitated, however, a feeder canal for the main section of the canal. This was very conveniently arranged for by the digging of a small channel from a bend on the adjacent North River, as the accompanying sketch map shows. This led to violent protests from the few local settlers but the idea was approved and the ditch was dug; it served well and can still be seen today.

It was not until 1833 that the Grenville Canal was far enough advanced to be used by Durham boats and not until the start of the 1834 navigation season that the three canals were officially opened for use. Any recital of all the reasons for the slow progress would be a tedious list of the problems of working in an isolated location amid untouched forests, under the extreme heat of summer and trying cold of spring and fall, with an outbreak of cholera to cap it all. But the canals were finished and were soon in regular use, despite the slowness of passage through them, and the limitations of the upper 3 locks on the Grenville Canal. All the other locks had dimensions not much different from those on the Rideau Canal (approximately 130 ft by 33 ft wide) but the upper three Grenville locks, having been built first—before the importance of steam navigation was fully realised—were only approximately 107 ft long and 19 ft 4 in wide. The 'approximately' is necessary since dimensions of the three locks differed slightly, those of the upper lock being the smallest, so that it was to be the bottleneck of the Ottawa-Rideau system, much to the continuing chagrin of the Duke of Wellington almost until his dying day. The canals were used for military purposes, even though the initial state of alarm gradually subsided, but they quickly came into regular commercial use for traffic up the Ottawa as settlement got under way and for traffic going through the Rideau Canal to Kingston and beyond, this being the principal water route to Lake Ontario until the completion of the St Lawrence Canals in 1855. Increasing amounts of sawn lumber passed through on specially designed barges, from the saw mills of Ottawa on their way to Montreal and even to markets in the United States. And passenger traffic between Ottawa and Montreal correspondingly increased until the coming of railways to the Ottawa Valley.

Two fleets of fine white paddle steamers provided the through service, one from Ottawa to Grenville, the other from Carillon to Montreal. A portage road initially provided a quicker passage between these two intermediate points than could sailing through the canals, but in 1840 a 12 mile portage railway was constructed and this served until through steamer service was given up in 1910. It was constructed to what was known as the provincial gauge (5 ft 6 in) and, because of its completely isolated position, it remained as the only broad gauge railway in North America after standard gauge (4 ft 8½ in) had become otherwise universal. The railway was privately owned and operated, for most of its life as a subsidiary of the river steamboat company, but the canals were always under public ownership, that of the Dominion of Canada from the time of Confederation in 1867.

Even before this time, it had become clear that canal reconstruction to more up-to-date standards was essential. This major project was carried out between 1873 and 1882, after which all locks were 200 ft long, 45 ft wide with a 9 ft depth over sills. Opportunity was then taken to revise the lay-out of the Carillon Canal, providing just two normal falling locks. The building of a crib dam across the Ottawa River between Carillon and Chute à Blondeau, essentially to assist the movement of timber down the river, flooded out the rapids at the latter location, thus making the single lock in the second canal unnecessary. The gates were removed but the excavated channel remained until all three canals were submerged beneath the water of the Ottawa River. A great dam just upstream of the entrance to the Carillon Canal was built by Hydro Quebec in 1963 for the generation in an integral power station of 840,000 hp. Remains of the upper Grenville and lower Carillon locks were fortunately preserved but the full lift of all three canals is now provided by a single modern lift lock, electrically operated, with about the same dimensions as the older locks but with a single lift of 65 ft.

Since the Ottawa River Canals were located about 60 miles east of Ottawa and west of Montreal respectively, and on the opposite side of the Ottawa River to the successive main roads between the two cities, they were relatively little known by Canadians apart only from local residents and those who used them. As with almost all other Canadian canals, freight traffic declined steadily during this century, although a regular service of small vessels carrying bulk oil from Montreal to Ottawa operated until replaced by a pipeline in 1953. As freight traffic decreased, however, passenger traffic increased and in the years just before their disappearance, the pleasant little canals on the Ottawa were carrying quite heavy traffic in pleasure vessels during the

Map E: The Ottawa River Canal

few weeks of high summer. Occasionally, they would have the unusual experience of passing large barges loaded with structural steel from the lower to the upper parts of the Ottawa, coming from a great fabricating plant at Lachine for use in bridges and similar structures in the Ottawa area. Now the canals have gone but passenger traffic continues to increase and the occasional load of special freight such as steel still comes up the river, all vessels now using the efficient Carillon lift lock, efficient but far different from the little locks so pleasantly operated for 130 years by the lock staffs on the quiet and peaceful Carillon and Grenville Canals which will always remain such a happy memory for those who knew them.[2]

The Rideau Canal

Similar in essentials, and a vital part of the Ottawa-Rideau system, is the Rideau Canal, built concurrently with the Ottawa River Canals but still serving today with its original locks still in use. It is carrying in high summer a greater volume of traffic than it has ever had in its long history. It is happily significant of changing times to find that a large proportion of the pleasure vessels that now sail the Rideau Canal are of United States registry, the crews often unaware that they are enjoying a canal constructed as a defence measure against possible invasion from their country. These visiting vessels are not merely from adjacent lake ports in New York State but sometimes from ports down the Mississippi as far as New Orleans, or from Hudson River ports coming to the Rideau by way of the New York State Barge Canal and its branch into Lake Ontario at Oswego. Vessels will come occasionally into the Rideau from Florida ports, having sailed up the Atlantic Intracoastal Waterway to New York, and at least one vessel has sailed on the canal into Ottawa from Portland, Oregon, having come down the Pacific coast, through the Panama Canal, across the Gulf of Mexico, up the Mississippi and then through the Great Lakes. It is becoming common practice for vessels to be sailed from Montreal up the Ottawa, along the Rideau Canal to Kingston and then down the St Lawrence Seaway back to Montreal, a modern version in power-operated (or assisted) pleasure craft of the former 'Triangle Tour' which the steamboat services at the turn of the century used to provide and advertise widely.

This pleasant modern usage of the canal is in great contrast to its beginnings, to which we must now return. The need for the alternative military and naval route between Montreal and Kingston was the imperative behind the Commission sent out in 1825 (see p66). Headed by Sir James Carmichael Smyth, it reported on 9 September 1825, recommending the completion of the Ottawa-Rideau canalisation project. The Duke of Wellington strongly urged upon the British Government adoption of the Commission's recommendations but, although it was agreed that the works should start, a sum of only £100,000 was proposed as that which Parliament should be asked to approve. The Duke was so anxious to maintain secrecy about these proposed defence works that he persuaded the Government to limit initial expenditures for the canals to £25,000 since they would come under the Colonial Office and so could be authorised by the Secretary of State without having to be referred to Parliament. This was done, but it was probably the root cause of the financial problems that arose as the Rideau Canal approached completion. With the several reports available to him, Wellington could have been under no illusions as to the magnitude and difficulty of constructing the Rideau Canal through more than 120 miles of virgin forest, but he placed every confidence in the officer of the Corps of Royal Engineers whom he selected to be the Commanding Officer, Lieutenant Colonel John By. His confidence was not misplaced. In a very special way, John By became one of the great early builders of Canada.

He landed at Quebec on 30 May 1826. After spending necessary time at the military headquarters there, he moved in August up to Montreal where he opened a temporary office. Within a month, he had made the somewhat difficult canoe journey up the Ottawa, passing the incomplete Grenville canal works on the way. The Governor, the Earl of Dalhousie, was with him. Together they selected the site for the start of the Rideau Canal in a natural gully, just opposite the fledgling village of Hull on the north shore of the river. This was a critical decision. The Rideau River falls into the Ottawa River a distance of about 40 ft over twin falls of unusual beauty (hence the name Rideau) so the canal had to use a different route up the escarpment provided by the high southern bank of the Ottawa at this location. The most desirable site, close to the great Chaudière Falls of the Ottawa, could not be used. The necessary land had been bought surreptitiously (even in those pioneer days), and then offered to the Government at an outrageous price which the governor refused to authorise. The chosen site was still satisfactory and it still provides a most picturesque setting for the northern end of the canal, right in the centre of the modern city of Ottawa.

There were other engineering problems to be dealt with without delay, the most important being the construction of a bridge across the Ottawa River to give access from the village of Hull to the uninhabited south bank where the canal works would be situated. John By laid out the general lines of this, the first bridge across the Ottawa, before he returned to Montreal and his active winter of planning and recruiting contractors for the major works which he already knew would be involved. He left a clerk-of-works in charge and work on the first masonry arches started in October. The bridge was a major structure with eight spans, the main one with a span of 200 ft which was bridged in 1827 with an unusual timber truss. Replaced in 1836 and again in 1872, this crossing of the Ottawa is still the main link between Ottawa and Hull, although now supplemented by other more modern bridges. John By moved up to the site of the works early in 1827, having decided to make his headquarters at the Ottawa end rather than at Kingston where he could have enjoyed the facilities of the military establishment, a decision that gives a clue to the sterling character of this great engineer-officer. He built a home for his wife and daughters close to the flight of locks up from the Ottawa. Close at hand were the workshops and stores. As settlers were attracted by the possibilities offered by the canal works, the beginnings of a small town were laid out, so well laid out that today the modern city of Ottawa is indebted to John By for the width and location of its main thoroughfares. At a dinner held when the works were in progress the little construction camp was named Bytown, in honour of the commandant, and this name continued in use until 1855 when 'Ottawa' was adopted for the growing town which was to become the capital city of the Dominion when it was formed in 1867.

It was not until the summer of 1827 that Colonel By was able to go right through to Kingston along the route of the canal, somewhat naturally in a canoe provided by the Hudson's Bay Company and manned by some of their most experienced *voyageurs*. If he was appalled by the magnitude of the task ahead of him, he gave no inkling in his written reports. Instead, he sent one of his young RE officers to make a more detailed examination of the route, on the basis of which greatly revised estimates of cost were prepared. Later in the year, he sent the same officer (Lieutenant Pooley) over to London with his estimates, clearly so that his emissary could speak from personal experience. Of special significance was the appeal that Pooley took to London for an enlargement of the size of the necessary locks. A towing path had been included in earlier general plans for the canal. It is a measure of By's vision that, even in 1827, he could see that steamboats would very soon take over all haulage in the canal and it was with this in mind that the larger locks were so urgently commended. Most fortunately, his arguments were accepted and the original dimensions of 100 ft by 22 ft with 5 ft draft changed to 134 ft by 33 ft with a normal draft over sills of 5½ ft. It can now be seen how unfortunate was the corresponding neglect of the smaller dimensions of the upper three Grenville locks which were to restrict traffic through the Ottawa-Rideau system so seriously for the next fifty years.

A considerable volume of excavation was necessary, as was also the construction of two large earthfill dams and four more locks (in two pairs) in order to get the canal back to the Rideau River after it left the head of the flight of the eight initial locks. These works now add greatly to the beauty of the centre of Canada's capital city, the canal within its boundaries being a unique park lined by gardens and pleasant driveways. Canal and river come together at a scenic spot known as the Hog's Back. Thereafter the canal is the Rideau River, suitably canalised, until the Rideau Lakes are reached. The level of some of these is also controlled by dams with appropriate locks leading the navigation channel to the headwaters of the Cataraqui River down which the descent to Kingston is made in similar fashion. It is easy today to see how convenient is the route that was selected (especially when viewed from the air) but it still beggars the imagination as to how Colonel By and his young assistants were able to select such admirable sites for dams and locks without the benefit of any preliminary surveys and with the surrounding forests coming down to water's edge. But they did and constructed a total of 46 masonry locks and 52 dams, all in five summer working seasons and at a total cost of about £800,000.

These bare statistics give some indication of the magnitude of the civil engineering construction project represented by the building of the Rideau Canal, one of the greatest of such projects carried out up to that time in North America and one which still commands the admiration of all who come to know it and have some appreciation of construction operations. So well was the masonry work done that the original locks are all still in use. They have naturally been well maintained down the years, with minor repairs and replacements as necessary, but to anyone with a sense of history it is a moving experience to be locked through one of the old structures and so to be able to examine at close hand the beautifully worked stones, or even to visit one or other of the old quarries from which the stones were obtained, in some of which unfinished blocks of rock may still be seen now surrounded again by forest. One of the dams, that at Jones Falls, is a true arched masonry dam 350 ft long at its crest and rising 60 ft above the bed of the stream that it dammed. It is a magnificent structure, although but little known since it is (perhaps fortunately) some miles from the nearest main road, a visit to the dam and associated locks necessitating a good walk but a walk that is well rewarded. Jones Falls is one of the real beauty spots on the Rideau Canal, its flight of three locks in a sylvan setting being one of the scenes along the canal long remembered after a first visit. There is another flight of three locks at the Kingston end of the canal, again in a beautiful setting, although here the presence of roads and a main railway line are constant reminders of the times in which we live and so tend to mitigate the historical feeling engendered by other more remote parts of the waterway.

Most of the masonry work was carried out under contract by experienced builders of Montreal, the names of some of whom are still carried by Montreal families and institutions. So well was Colonel By pleased with their work that he presented four of the builders with suitably engraved silver cups. He encountered almost an equal degree of difficulty with his contracts for earthworks since these attracted the inexperienced. Just as was the case with the Ottawa River Canals, land owners made life difficult for the commanding officer, on top of everything else, untouched land in virgin forests suddenly acquiring (in their eyes) unusual value for which recompense was demanded, sometimes in the courts with consequent frustrating waste of time for the busy engineers. The heat of summer was again a more serious problem than the cold of winter, 'swamp fever' (probably a form of malaria) causing the death of many of the canal labourers and almost taking the life of John By himself on one occa-

sion. All the difficulties were successfully overcome, however, and the canal was substantially completed in the fall of
1831. It was not until 24 May 1832 that Colonel By with his
family, some friends and associates, started on the first
through passage of the canal from Kingston, steaming into

Map F: The Rideau Canal

Bytown on 29th of the same month. One would imagine that the man who directed so successfully this vast enterprise in the forests of Canada for the defence of this outpost of Great Britain would have been suitably honoured by his country. He was not, but subjected rather to a Committee of Enquiry into the canal finances since Treasury regulations had not been strictly followed. It is a tangled tale; suffice to say that he was exonerated completely but, I feel certain, died of a broken heart at Frant in Sussex on 1 February 1836 less than four years after that triumphant voyage from Kingston to Bytown.

The canal did carry military supplies to Kingston; small naval vessels did sail through on their way from Montreal to Kingston. Almost from the first, however, it served mainly as a commercial waterway, providing also a vital communication channel for the opening up of the country through which it passed. Timber was a major item of freight throughout most of last century. Coal was carried for the railways when they came into the Rideau country. Cheese boats were a familiar kind of special vessel, the size of the freight sheds on the canal in Ottawa (dismantled only in the 1960s) being testimony to the varied and extensive freight traffic on the canal in its heyday. Small specially built passengers vessels provided a convenient and pleasant service between Ottawa and Kingston in those happy days when 'travel by water' was an accustomed means of moving about on business or on pleasure. It is pleasure travel today that alone makes use of the Rideau Canal, and to a steadily increasing degree. So busy have some sections become that one lock (at Newboro) has already been changed over to electrical operation. The change has been well done without interfering too much with the historic appearance of the lock but there has been inevitably some public concern at the possibility of further 'modernisation' spoiling the unique character of the Rideau Waterway. Special studies are in progress even as these words are being written as to the future development of the canal, so that its continued use is well assured.

Some indication of what a journey through the canal is like must be given even though limitations of space mean that a brief and superficial account only is possible. After the lovely sail through the garden-flanked section in Ottawa, following the interesting lift up the initial flight of 8 locks beneath the Parliament Buildings of Canada, a winding section of the Rideau River is followed beyond the Hog's Back locks and dam with one lock to be passed before Long Island is reached. Here is one of the fine major masonry dams and a flight of 3 locks seemingly 'tucked into' the east bank of the River. Then follows the 'Long Reach', a stretch of clear sailing for 24 miles through pastoral country. Four locks have then to be passed in quick succession before Merrickville is reached with its flight of 3 locks and its interesting block house, built as a defence measure in case the canal was attacked by marauders coming across the St Lawrence. Four more locks follow, all in pleasant settings, before the important railway centre of Smith's Falls is reached, again with a flight of 3 locks in a park-like setting in the centre of town. Two more locks, a singularly beautifully situated one at Poonamalie, bring us into Big Rideau Lake, headwaters the Rideau River, from which we pass into Upper Rideau Lake. This entry is accomplished by use of the Narrows lock, near to which stands another block house. Having risen through a total of 33 locks, we have now arrived at our summit level. It is 277 feet above the level of the Ottawa River.

I wish that I could pause to attempt to describe the beauty of this lakeland but we must push on, into Newboro, Clear and Indian Lakes, past Chaffey's lock, thought by some to be the loveliest of all, across Opinicon and Sand Lakes and so to the spectacular flight of locks and dam at Jones Falls. We pass only the crest of the dam so that disembarking is necessary if the dam is to be seen properly, as it should be by all voyagers, and we can do this by stopping at the wharf of the century-old fishing hotel at the foot of the flight of locks, still operated by the Kenny family. We are now in the headwaters of the Cataraqui River, and the scenery from here on differs slightly from that we have been enjoying, but is still delightful with attractions all its own. Locks at Brewer's Mills flooded a great swamp, Cranberry Bog, where Colonel By almost lost his life. They lead us, after one more lock passage at Washburn's, into the water impounded by the dam at Kingston Mills. The three locks here lower us to the level of Kingston Harbour, a drop of 162 ft from the level of Upper Rideau Lake, achieved through 14 locks. And so we sail into the harbour of Kingston, past the reconstructed Fort Henry, now a fine tourist attraction, past the Royal Military College of Canada on its fine site, once part of the naval dockyard of earlier days, and so into Lake Ontario.

It is difficult to associate this pleasant journey of today with the stern defence measures for which the canal was constructed. One of the Duke of Wellington's sons (Charles) sailed through the canal and duly reported his experience to his father. One wonders what he said about the quiet beauty of the route he had followed, feeling some regret that the great Duke did not himself have this pleasure in view of his long-time attention to its building. It met his defence requirement but it has given the Canada of today a waterway of real beauty, a silver chain of rivers and lakes which gives delight to those who journey on it by water, all who visit its shores, and the summer residents of the innumerable summer cottages that have been built on its shores and on the many islands that enliven its waters.[3]

CHAPTER 5

Canals Along Indian Routes
The Trent Canal System
and the Murray Canal

The Trent Canal, although not far from the Rideau Canal (in terms of Canadian distances), differs from it in many respects. It links Georgian Bay, a part of Lake Huron, with Lake Ontario. Its entry into the latter is about 50 miles west of Kingston, the southern end of the Rideau system. The Trent Canal was not constructed, however, for defence; different sections of it were built to aid early settlers with the movements of their produce (especially lumber) and for other commercial purposes. It was at one time under the jurisdiction of the provincial government of Ontario, being taken over by the Government of Canada, through its Department of Railways and Canals in 1892. Even today, one still cannot sail through from Lake Ontario to Lake Huron without having to use two lift locks (vertical lifts) and one of the two marine railways that used to distinguish the Trent system, the other having been replaced by a canal lock within relatively recent years. Its gradual development was not, therefore, the responsibility of one man as was the Rideau Canal; many engineers have contributed to its intermittent construction. To associate the Rideau and the Trent Canals, therefore, as has been done in one or two recent reports, is a purely artificial device having no real basis in fact. They are two quite separate and distinct canals, equally lovely, but of very different historical significance, each serving today the needs of pleasure boating to a surprising extent. And the Trent Canal has the added distinction of

81

being closely associated with the Murray Canal, one of the least known of all Canadian canals but one with more political overtones to its construction than would seem possible.

Reference to the accompanying map will be necessary in order to appreciate even the main features of the meandering course followed by the Trent system. It uses, again, two rivers and a connection between them provided by a smaller river and a chain of lakes—the River Trent flowing into Lake Ontario at the town of Trenton, and the Severn River which connects Lake Simcoe (one of Ontario's larger inland lakes) with Georgian Bay in Lake Huron. It will be seen that the Trent follows a tortuous course from its source in Rice Lake. Beyond this the canal has to use the smaller Otonabee River as a connection with the chain of lakes, through which its course runs generally westward as far as Lake Simcoe. As it follows the Severn River out of Lake Simcoe it goes first to the north, finally swinging west again for its final section into Lake Huron. It would require a very large and detailed map indeed to show the almost innumerable lakes through which the canal threads its way. Almost the whole route of the Trent Canal lies within the Precambrian Shield, that vast area of glaciated ancient rocks that makes up almost half the land area of Canada. The area is distinguished in most parts by the intricate network of interconnecting streams and lakes which makes its physiography so pleasant and delightful in summer in southern Canada but which renders travel through it hazardous and exacting to the uninitiated. The retreat of the last glacier from this part of the Shield took place about 10,000 years ago. The resulting glacial drainage is responsible for many of the river courses of today, and the links between adjoining river basins such as are used by the Trent and Rideau Canals. Geology is, therefore, the determinant of the unusual course followed by the former, the exact route between Georgian Bay and Lake Ontario through the Kawartha Lakes having been determined long ago by the first Canadians, the Indians.

Complex though the route appears to be at first sight, it

had gradually been determined by the Indians as the most convenient passage through this beautiful lake country, long before the coming of the white man to North America. It may have been followed by a young Frenchman sent up the Ottawa by Champlain in advance of his own pioneer voyage up the Grand River (as the Ottawa used to be called) but no record of this was made. Champlain, however, as with all his major journeys, wrote an account of his first and second journeys up the Ottawa and had these published in Paris. His first journey, as far as Pembroke, was in 1613. Two years later, he went again up the Ottawa, this time right through to Georgian Bay (see p 2) where he joined Huron Indians for an attack on their inveterate enemies, the Iroquois, as he had promised them at the time of his first journey. The Iroquois lived in New York State. To reach them, the Hurons took Champlain by canoe along the route followed today by the Trent Canal. They did battle with their enemies but even Champlain's help and encouragement did not ensure a victory. Retreat was necessary, along the same route once Lake Ontario was crossed, and Champlain had to be carried 'in a basket' for much of the way since he had been wounded in the fight. It was a journey he would never forget. The Indians would not allow him to return by way of the St Lawrence, of which he had heard. He was taken back to the Huron encampment on Georgian Bay where he had to spend the winter, returning safely to his pioneer headquarters at Quebec City only in May 1616. From Champlain, therefore, we have the first account of this old Indian route, now followed by the Trent Canal. It is not surprising that valuable Indian artifacts have been discovered at a number of places along the canal route, yet another feature of unusual interest of the area served by the canal.

For almost two hundred years after Champlain's journey, this 240 mile 'portage' (if the old word may be used in this unaccustomed way) between Lake Huron and Ontario remained as a winding waterway through the forests, used mainly by Indians since it was not until the start of the nineteenth century that settlement of the area through which the canal now runs started in earnest. Trenton (or Trent-Port as it was first known) had its start at this time when great quantities of timber were brought down the Trent River to the lake, on their way to Montreal and Quebec for the markets of Europe. The town of Peterborough was first laid out in 1826. Although this inland town was closer to Port Hope than Trenton, it was to the latter that lumber from the Rice Lake and Peterborough area had to go, there being no roads at that time, water transport being essential for getting logs down to the lake. It was, therefore, natural that agitation for a canal to improve the Trent River should start. The first survey appears to have been made in 1833 by a Mr Baird who prepared plans for a canalised routed through to Lake Simcoe. So great was the estimated cost that revised plans called for a combined canal and railway; it is strange to think that this combination is still in use today. Locks were started near Peterborough and at the exit from Rice Lake but provincial funds had to be diverted from this canal work in view of the disturbances of 1837. Work was suspended until after the union of Upper and Lower Canada in 1841 when the newly created Board of Works arranged for the resumption of work. The two locks mentioned were completed by 1844 and gave a 53-mile through waterway from near Peterborough to Healey Falls but still without the essential connection to Trenton and Lake Ontario.[1]

At about the same time another lock was being built at Bobcaygeon to facilitate passage between Sturgeon Lake and Cameron Lake. This was at the behest of early settlers and lumber 'barons' who were then reaping the harvest presented by the untouched forests around. So isolated were some of the settlers that one of them has recorded that 'the insurrection (of 1837) was suppressed and tranquility restored before we heard of its interruption'. And from the log cabin in which these words were written 'John had to walk seventy or eighty miles to give his vote' in a general election.[2] Politics were taken seriously in those days, as we are so interestingly told in some of the classic tales of pioneer life in the forests of Canada that were written in this area. The construction of every small lock was a political adventure! A second one was built at Lindsay on the Scugog River which was eventually finished in 1843, thus linking Scugog Lake with Sturgeon and three other lakes. These isolated sections of what was to be the Trent Canal eventually did assist with the shipment of timber but only with the help of short railway lines coming up from the shores of Lake Ontario, fortunately so since no water outlet to the lake was yet available. There was justifiable agitation for completion of the canal as the timber business reached its peak about mid-century.

Haphazard construction of individual locks and associated timber slides continued, all on the right route but still not giving a through connection. It would be invidious to recite the long drawn out negotiations that surrounded further development of the main canal project but mention must be made of the local public outcry that greeted an announcement of the Liberal administration in Ottawa of Alexander Mackenzie placing the whole Trent Canal project under the provincial government of Ontario, despite the fact that the British North America Act (the Charter of the Dominion of Canada) clearly placed all canals under the administration of the Government of Canada. 'Trent Canal Associations' were formed. Public meetings were held at many of the small settlements along the route. Newspaper editorials fulminated about the need for a clear route through from Lake Ontario to Lake Huron. Ontario did complete two more locks that linked together all the Kawartha Lakes on the canal route. Then in 1887 Sir John A. Macdonald, now again the Conservative Prime Minister of Canada, appointed yet another commission to investigate the proposals for the Trent Canal. It reported favourably in 1890 and by 1896 contracts were awarded for the completion by the Government of Canada of the works necessary to give a through route from Trenton to Georgian Bay, even if not a through canal.

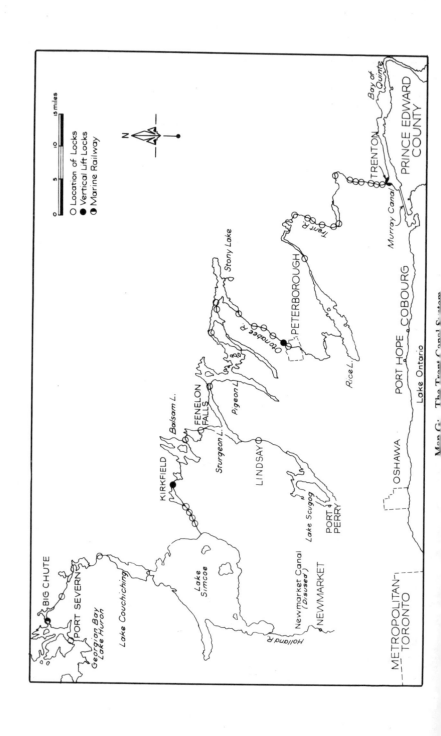

Map G: The Trent Canal System

By the year 1907, through navigation was possible from Healey Falls on the Trent River to Lake Simcoe. In succeeding years, older works were rebuilt and the remaining works necessary for completion of the whole canal were completed, thus giving 8 ft navigation from Lake Ontario as far as Peterborough and thence 6 ft navigation as far as Swift Rapids on the Severn River and for the final 8 miles from Big Chute into Georgian Bay. Marine railways were installed at these two falls on the Severn River, and for the 8 miles between them a draft of only 4 ft was possible. The first vessel to make the through trip was a motor launch named *Irene*. She sailed out of Trenton on 3 July 1920, arriving in Port Severn at the Lake Huron end of the canal on 12 July. It is to be noted that this was almost 86 years after Colonel By first sailed through the Rideau Canal, further evidence of the profound difference between the histories of these two adjacent systems.

The remaining marine railway at Big Chute naturally continues to attract much attention from visitors to the Trent system. It still gives good service after its fifty years existence because of the good maintenance it has received. In a simple timber 'cradle' running on steel rails, even a 20-ton cruiser can readily be hauled up the normal lift of 58 ft, the extent of the lift being in itself an indication of the picturesque location of the railway. Nearby is the Big Chute water power plant, built as early as 1908 by the Simcoe Light and Power Company but since 1914 a part of the extensive system of the publicly owned Ontario Hydro. Proposals have been made for the replacement of the marine railway at Big Chute by the construction of a new lock just as was done at Swift Rapids. Fortunately this has not yet been put in hand and there are many who hope that it will always remain just a plan on paper so that this unique feature of the Trent Canal may long remain in service. The real objection to the building of a lock, however, is biological since it is feared that, if a through water connection was provided, sea lamprey from Lake Huron would gain access to Lake Simcoe with possibly devastating effects upon the fishing there that is so widely enjoyed. Certainly until the lamprey problem (a very serious matter in general for Great Lakes fisheries) is solved, travellers on the Trent system will continue to have the pleasure of seeing their vessels transferred high and dry from the one section of the Severn River to the other, up the steep incline at Big Chute.

The two lift locks or vertical lifts on the canal serve a similar purpose, having been constructed as the most con-venient and economical means of achieving the big lifts necessary (because of the lie of the land) at Peterborough and Kirkfield. Summit level of the canal is on Balsam Lake which, with its connecting waters, is at an elevation of 841 ft above mean sea level or 596 ft above the level of Lake Ontario at Trenton. A drop of 123 ft is necessary before the canal reaches Lake Simcoe and 49 ft of this total is achieved through the twin lift lock at Kirkfield. This is essentially a structural steel installation, opened in 1907, the operation of which is identical in principle with that of the larger lift lock at Peterborough which may, therefore, be described in more detail.

Swift rapids on the Otonabee River in the vicinity of Peterborough made it desirable to construct an artificial canal around the eastern limits of the city, the topography making readily possible the concentration of fall in one loca-tion. The normal lift is 65 ft which would have required several ordinary locks with mitre gates whereas by the adop-tion of the vertical lift principle, the entire rise (or fall) takes place at the one location. As can be seen from an accom-panying illustration, the main towers are of concrete; they make an imposing sight. The lock was started in 1896 and finished in 1904. It was operated quite satisfactorily for 59 years with only a minimum of maintenance but with diligent care in its operation. In 1963 the structural steel of the locks and supports was renovated and in 1964 the hydraulic, mechanical and electrical parts of the installation were given a major renovation, the opportunity being taken to provide automatic electrical controls for operation replacing the original manual control with a new control cabin located at the edge of the fixed approaches to the locks. The cabin of the original installation was at the top of the central tower, giving the operator a remarkable if eerie vantage point from which to check on his operation. This is very simple in principle, the chamber that is to descend being filled with rather more water than that which has to come up; the excess water is so gauged that its weight is sufficient to move the two locks, down and up respectively, under the control of fluid in the hydraulic pistons on which the locks ride. After the gates at the ends of the locks have been closed and locked in position, the simple opening of a connecting valve on the hydraulic system allows the operating fluid under pressure to move from one 'presswell' (an expressive name) to the other and the move commences. The two pistons are installed in wells 75 ft deep and 16½ ft in diameter. Adding interest to the whole scene is a tunnel which carries a single road, of some importance, immediately beneath the approach to the locks.

The lock chambers or caissons are 140 ft long and 33 ft wide, the extra draft of water necessary in one to ensure movement being only 12 in. Dimensions of locks vary throughout the Trent system; this is not too surprising in view of the manner in which it was built. The first 18 locks at the Lake Ontario end have a clear length of 154 ft with a width of 33 ft, the latter being standard for most of the canal. After the lift lock at Peterborough, the clear length within locks is 120 ft as far as the Severn River where the Big Chute railway severely limits dimensions. The final lock at Port Severn, giving into Lake Huron, has clear length of only 84 ft and a width of 25 ft. This is lock No 49. The lift lock at Kirkfield is No 36; that at Peterborough No 21. These numbers will indicate generally the distribution of locks along the canal. The drop from Lake Simcoe into Georgian Bay is normally about 138 ft, the water in the bay being about 335 ft above the level of Lake Ontario. On the St Lawrence route most of this represents the drop over Niagara Falls and in the

Niagara River. Sailing the 240 miles of the Trent Canal is, therefore, quite an adventure with its rise of almost 600 ft from Trenton to the summit at Balsam Lake made by means of 35 locks, two of which (at Healey Falls) have a combined lift of 54 ft. The descent to Lake Simcoe and thence into Georgian Bay is not quite so spectacular from the point of view of lockage but this western section presents some of the most beautiful scenery along the route, although the entire canal route is quite outstanding in its lovely surroundings.[3]

This applies even to the stretches of artificial cutting that were necessary to give some of the convenient connections between lakes and access to suitable lock sites, as also at Peterborough, the total length of man-made canal being no less than 33 miles. The combined watersheds of the Trent and Severn River systems amount to 7,200 square miles (considerably greater than the area of Yorkshire). The many lakes in this forested area permit a high degree of water control so that along the canal there are many dams which serve as control dams for power plants as well as impounding dams for the canal. Wherever possible, power is generated from falling water in the vicinity of the canal. Approximately 100,000 hp is thus generated and fed into the Ontario Hydro system as yet another contribution of the Kawartha Lakes region. So critical are the depths of water in some sections of the canal, because water levels must be controlled for the water power plants, that any vessel drawing more than 6 ft must still give canal authorities twelve hours notice before entering the Lake Ontario end at Trenton. I must refrain from quoting any more statistics lest they give the idea that the Trent Canal is one only to be enjoyed by engineers. It is, on the contrary, another lovely waterway that is giving great pleasure to an increasing number of visitors during summer months, its western section reputedly carrying the largest number of pleasure craft of any comparable waterway in this part of north-eastern America. Many of the lakes it serves are renowned for their fishing potential, and the innumerable islands in them and their many isolated rocky shores provide sites for a large number of summer 'cottages' (some of them minor palaces) where once stood the log cabins of the pioneers lonely in the woods.

Reference to the map on page 85 will show that the port of Trenton is located at the head of an unusually well sheltered bay, that of Quinte. The large peninsula that forms this bay is one of Ontario's historic and justly renowned areas, Prince Edward County. Its significance for all water travel in the eastern part of Lake Ontario will be obvious from the map. Those sailing eastward along the north shore of the lake were naturally attracted to the possibility of shortening their journey appreciably by using a portage across the narrow neck of the peninsula which is just three miles south of Trenton. So obvious a portage route is this that it was well used by Indian travellers long before the days of the white man's coming to

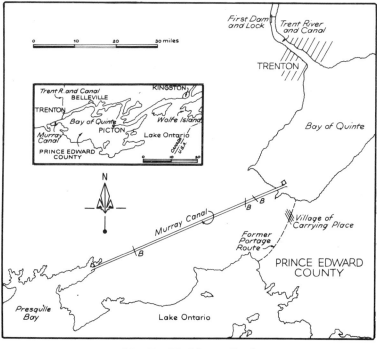

Map H: The Murray Canal

these parts. French-speaking and then English-speaking travellers of earlier days followed the Indian practice, the use of the portage becoming so well established that the small settlement that grew up around it was, and still is, called Carrying Place. Here was an obvious location for another 'cut-off canal' and agitation for this desirable improvement to navigation into the Bay of Quinte started well before the end of the nineteenth century. Eventually a canal was built, started in 1882 and completed in 1889. But was it built at the narrow neck so long used as the portage route? Not at all, but on a parallel alignment about one mile to the north. Instead, therefore, of being a mere 1½ miles long, the Murray Canal as built has a total length between eastern and western piers of no less than 5.15 miles, distance between the entrances to the two dredged approach channels being 7.53 miles of which 6.80 miles are in a straight line. Since there are no locks, navigation through the canal is relatively easy, its top width being never less than 124 ft with a clear draft of at least 9 ft 6 in at normal low water level. Three swing bridges cross this pleasant quiet channel, two for roads and one for the Canadian National Railways' line that serves the town and port of Picton. So well established does the Murray Canal now look in its rural setting that few ever question why it is where it is and not at Carrying Place. Readers of this book will want to know, however, so it must be admitted that according to popular local belief, for political purposes that can best be imagined, it had to be located in the constituency of Northumberland and not in that of Prince Edward County. By chance at the time of its construction, the two members representing these constituencies belonged to opposing political parties, the member for Northumberland being a supporter of the Conservative Government then in power.[4]

CHAPTER 6

Some Minor Canals

The major navigation canals of Canada, with the exception of those constituting the St Lawrence system, have now been described. Their histories here so briefly summarised present the same pattern of high hopes, thwarted ambitions and keen disappointments that characterised so much canal development in all developed countries. Railways were late in coming into their stride in Canada and so the 'canal fever' persisted later in the nineteenth century than, for example, in Great Britain. Great hopes were centred not only on the canals that have been so far described but on many smaller projects throughout eastern Canada and even, to a limited degree, also in the west of Canada. In some cases the hopes evaporated, leaving no trace behind except in long-forgotten papers. In others, work was sometimes started and sometimes completed. There are still to be found, therefore, small isolated canal or lockage projects throughout eastern Canada, a few still in use, others now abandoned. This record of the canals of Canada would not be complete without some reference to these minor projects. Any treatment must naturally be selective but in the following few pages the attempt is made to give some indication of the more interesting and more significant of these ventures, interest—as will be seen—being not always related to size or location. The Ottawa River and its projected canalisation alone requires rather more detailed description and so it will be dealt with in the concluding chapter of this first part of this book, providing a fitting introduction to Part Two.

Branch Canals

Branches, from main canals that had proved successful, to communities which thought they would benefit from such water connections, provide an obvious starting point. Both the Rideau and Trent Canals had such branches. That on the Rideau takes off from Big Rideau Lake, near the summit level of the canal (in Upper Rideau Lake) and runs to the lovely country town of Perth. Known as the Tay Branch, this canal is just over 6 miles long. Originally promoted by a private company, after the success of the Rideau Canal was assured, work started in 1832, 4 small wooden locks being necessary and a certain amount of dredging. Limited though this feeder canal was in the size of vessel it could serve, it did assist with the growth of Perth and its early industries but it was never a commercial success. Maintenance became a problem and so eventually the Government of Canada took over the whole enterprise, rebuilding the locks, or rather building two new good masonry locks at Beveridges, near Rideau Lake. Small though the Tay Branch was, it seemed to be permanently involved in local politics. Even today, long after it has been used by any but very small craft, one can still hear it called 'Haggart's Ditch', the Honourable J.C. Haggart having been the Minister of Railways and Canals in the Government of Canada when it took over this small canal.[1]

One of the early locks on the Trent system was built at the town of Lindsay (see page 85) but this was then merely a local improvement to assist with lumber movements into the Scugog River and so into Lake Scugog. As the surrounding area became well settled, the possibility of using this lock as a link in a water route between Port Perry at the south end of Lake Scugog and the Trent system was recognised. The lock was rebuilt and some dredging done, the Port Perry Branch Canal thus being established. And it is still in use, although now for the ubiquitous pleasure craft that are to be found all over the Trent system in high summer. Of special significance, both in earlier days and certainly today, is the fact

Page 95. (Above) *The upper lock and pool at Kingston Mills, the southern end of the Rideau canal where it joins Lake Ontario at the mouth of the Cataraqui river. In the background is one of the original blockhouses built for the defence of the canal locks against possible attack from USA; (below) The flight of three locks at Jones Falls on the Rideau canal; they lead to a basin from which a fourth lock completes the total rise here of 58.5 feet. This rise is achieved by the impounding of the Cataraqui river by a large arched masonry dam*

Page 96. Trent canal: the lift lock at Peterborough on the official opening day, 9 July 1904. The lock looks much the same today, following rehabilitation of its operating machinery

Page 97. (Above) *Trent canal: opening day, 1965, of the concrete lock at Swift Rapids that replaces an earlier marine railway similar to that still operated at Big Chute* (below) *Houseboat in the lift lock of the Trent Waterway at Kirkfield, Ontario*

Page 98. *Trent canal: Boat on the marine railway, Big Chute, Ontario*

Page 99. *St Andrews lock on the Red river, Manitoba, close to the mouth of the river in Lake Winnipeg; the control dam assists in river regulation and provides support for a main highway*

Page 100. *Port Carling lock in the Muskoka Lakes, north of Toronto, in 1942: S S Segwun passes from Lake Joseph to Lake Rosseau*

Page 101.
A photograph taken probably in the 1860s near St Catharines shows one of the 27 locks and lock basins of the Second Welland canal

Page 102. *Third Welland canal: the Arthur of Toronto in one of the locks near the top of the Niagara escarpment, near St Catharines, in 1904*

that Port Perry is only about 30 miles from the great metropolitan area of Toronto with its more than two million inhabitants, access to the delights of the Trent Canal being thus most convenient.[2]

Poupore Lock

It will be recalled that the Rideau Canal starts at its northern end near the Chaudière Falls on the Ottawa River. The Ottawa is fed by a number of tributaries, the larger of which are fine rivers in themselves. One of these is the Lièvre River which from its source in the Precambrian Shield to the north joins the main river about 20 miles downstream of the Chaudière Falls and so the city of Ottawa. The Lièvre has always been an important logging river, so much so that at the great falls that occur about 3 miles up from its mouth there has grown up the thriving industrial town of Buckingham with associated power plants, paper-making and other industrial works. Logs from the forests have been brought down the Lièvre almost from the start of settlement on the Ottawa. There was, fortunately, a stretch of 24 miles of river just above the falls at Buckingham with only one impediment to smooth navigation for the tugs and other craft used in logging operations, this being a small rapid at Poupore almost at the midpoint of this stretch of river. Possibly because of the fine examples provided on the Rideau Canal 'just across the river', a lock was proposed as an improvement at the Poupore Rapids and was built between 1887 and 1895 by the Department of Public Works of Canada. It is a fine structure of masonry, the lock being 160 ft long and 32 ft wide, with 8 ft of water available over its sills for a total lift of 6½ ft. It was operated by the Department even though the'navigation' it aided was naturally confined to tugs and other vessels used for logging. During recent years, traffic consisted merely of two tow boats in the spring and fall. It is small wonder, therefore, that the entire facility was sold in 1956 to the James MacLaren Company (the great local paper enterprise) which almost immediately utilised the east side of the fine masonry structure as a foundation for one of its slash mills. It is a melancholy sight today to see this fine canal lock used for such a mundane purpose, blocked with good but utilitarian concrete.[3]

Locks in the Muskoka Lakes

The only Canadian canals now operated by a provincial government are three small locks and approach channels that are under the control of the Ontario Ministry of Natural Resources. They are located in the Muskoka Lakes region, a singularly lovely area that is conveniently located about 100 miles north of metropolitan Toronto. It has therefore become a favourite and well patronised holiday area. Traffic through these locks consists now almost wholly of pleasure craft even though they were built originally to assist with the movement of floating timber and for the small steamers that served the early settlements in the region before roads and railways had penetrated it.

A lock in the channel connecting Lakes Rousseau and Joseph, the Indian River, was the first to be built. Constructed in 1871, its walls were formed of rock-filled timber cribs following the practice of the time. Its immediate success is shown by the fact that a smaller parallel lock was constructed shortly after the first lock went into service, located 50 ft upstream. This smaller lock was substantially reconstructed in 1962-63; it is 83 ft long, 12 ft wide with a depth of 8 ft 6 in over sills. The main lock at Port Carling, as the settlement around the locks is known, had major repairs carried out in 1921 when a new steel highway swing-bridge was installed. Further deterioration was so rapid that an entirely new concrete lock with electrically operated steel gates was constructed to replace it, about 200 ft away, in 1952-53. It is 175 ft 6 in long, 33 ft wide with 8 ft 6 in over sills, dimensions that differ only slightly from those of the original lock, The maximum difference of level between the two lakes is only 4 ft, being even lower than this in high summer, but a lock is essential to obviate the swift water that even this fall created between the lakes.

The lock near Huntsville, between Fairy Lake and Mary Lake, was the next to be constructed, in 1874. It created a through waterway of 15 miles between the end of the pioneer road that reached the south end of Mary Lake and the new settlement of Huntsville, now an important rural town well served by road and railway. The lock measures 88 ft 6 in long by 24 ft wide and has an 8 ft lift, but its width was reduced to 20 ft in a major reconstruction carried out in 1946-49, when the lock was closed for three summer seasons. Concrete facing was applied to the old timber crib structure as a remedial measure.

Vessels Passing Locks at

Year	Port Carling	Huntsville	Magnetawan
1961	11,058	1,288	370
1972	23,348	3,810	1,023

The third Muskoka lock is at Magnetawan, 35 miles to the north-west of the Huntsville lock but still in the lovely Muskoka area. The lock was built between Ahmic and Cecebe Lakes. It is 112 ft long and 28 ft wide but with only 5 ft 6 in over sills, built originally of the then standard rock-filled timber cribs. It was reconstructed in rubble masonry in 1922. It was much used in lumbering operations in earlier days but it did also serve the needs of small passenger steamers which were thus able to sail 12 miles further up the Magnetawan River as far as Burk's Falls, a name suggestive of the original rugged character of this area.[4]

Despite the disappearance of all the original traffic through these small inter-lake locks, they are busier today than ever in their history. This is well shown by the increase in the number of lockages during the last dozen years, almost all being of privately-owned pleasure craft. The lock at Port

Carling did serve the fine fleet of small steamers that provided a convenient passenger service to the many summer
resorts on Lakes Joseph and Rousseau until the years immediately following World War II when the increase in automobile traffic had its usual dire effect.

St Andrew's Lock, Manitoba

Another lock still happily in use is far removed from all the
canal works so far discussed. The Red river of Manitoba is
one of the historic waterways of western Canada. Rising in
the United States, it flows due north through Manitoba,
entering Lake Winnipeg about 10 miles north of the town of
Selkirk which, in turn, is about 15 miles downstream of
Winnipeg, the great prairie city that is at once the capital of
the province of Manitoba and the 'gateway to the West'.
Once railways had come into the mid-west of the United
States, the Red River came into its own as the entrance to
the north and west of Canada, in place of the long and
arduous journey by lakes and rivers from Lake Superior.
The first steamboat on the Red River was prefabricated,
taken over the prairie in pieces and re-erected on the banks
of the Red. It brought, in turn, the first steam locomotive
down the Red River to the vicinity of Winnipeg when railroad building started on the Canadian prairies. River traffic,
therefore, has always been important on the Red River, at
first down from the Dakotas but, as development in Manitoba gradually extended northwards, eventually down from
Winnipeg and into Lake Winnipeg with its connections with
other waterways to the north and the west.

Flow down the Red River varies greatly. It has been the
cause of catastrophic flooding at Winnipeg on three known
occasions, as one result of which a great floodway has been
excavated around the eastern limits of the city. Navigation,
especially on the lower reaches of the river, was naturally
interfered with by varying river water levels. Accordingly, a
combined facility including a control dam, a navigation lock,
a fishway, and a highway bridge was built near Selkirk be-

tween the years 1900 and 1916, again by the Department of Public Works of Canada. Known as the St Andrew's Lock, this improvement to navigation was constructed during the first five years of this period, the highway bridge being completed only in 1916. Dimensions of the lock are 215 ft long by 46 ft wide with 10 ft of water over the sills. It has always been operated by the federal Department and although traffic through it has naturally varied a good deal through the years, its current use sees about 1,500 pleasure craft up to 40 ft long locked through every summer as well as 300 passages of tugs, barges, and longer pleasure craft which have included excellent passenger vessels giving valued service to the small points around Lake Winnipeg.[5]

The Baillie-Grohman Canal, British Columbia

Far removed as is the St Andrew's Lock from all other canals of Canada, there is a yet more distant canal project that must be mentioned if only to explain a name that will be found on maps of Alberta and British Columbia, almost on the boundary between these two far western provinces. It is on the western side of the Rocky Mountains (the most easterly of the three great ranges that make up the mountains of western Canada, often thought of erroneously as all being 'The Rockies') and on the main road running south from the Trans-Canada Highway at Golden to Kimberley and Cranbrook. Here, amid the fascinating scenery of the mountains, is to be found one of Canada's most unusual canal projects. If he stops at the small town named Canal Flats, the observant traveller may see this attractive sign erected by the Department of Recreation and Conservation of British Columbia:

A DREAM FULFILLED

It was the dream, in the 1880s, of W.A. Baillie-Grohman, British sportsman and financier, to reclaim these fertile lands from the annual river floods. His canal at Canal

> Flats diverted part of the Kootenay into the
> Columbia but was abandoned. The first
> successful reclamation was in 1883. Now
> 25,110 acres lie secure behind 53 miles of
> dykes.

As can be well imagined, behind these few words lies a story of high hopes and real adventure when the West was young. It so happens that a combination of local geology and hydrological factors has led to the headwaters of the great Columbia River being at this point only about one mile away from the upper reach of the Kootenay River. The intervening land is generally flat but the Columbia is normally 6 ft lower in elevation than the Kootenay. This unusual natural feature was noticed by William A. Baillie-Grohman when big-game hunting in this region. He immediately had the idea of excavating a ditch across the flats and so diverting the Kootenay into the Columbia. Prevention of flooding on the Kootenay (still a problem) was probably one motivation. Most fortunately, his proposal was vetoed. Rivers being what they are, one wonders what would have been the effect on this lovely area, and on the two rivers, had the scheme been implemented. Not to be outdone, Baillie-Grohman then decided to build a navigation canal between the two rivers, both of which were used in those early days (before the coming of the railway) as regular transportation routes.

Begun in 1887, the very year in which the Canadian Pacific Railway was completed as Canada's first transcontinental railway, the canal was completed by 1889. Most of the excavation was done by scrapers and by Chinese labourers (many of whom were brought to Canada for work on the early railways) using simple wheelbarrows. One lock was naturally necessary. This was constructed of wood and was 30 ft wide and 100 ft long. A wooden guard lock was also constructed at the upper (Kootenay) end of the canal—all this as a private venture. Hoped-for traffic failed to materialise, the records suggesting that only two major steamers passed through, one in 1894 and the other in 1902.

The latter, the *North Star*, was too wide and too long to fit in the lock so the timber gates were destroyed by burning. The upstream gate was first removed, the vessel locked in, a temporary dam being built around its stern. The lower gate was then removed after another dam had been built, this being the first of the two to be removed by blasting. Remains of the lock as well as the canal excavation may still be seen, for this tale is no fabrication but a brief record of one of the most unusual canal ventures in the history of Canada. The owner returned to his castle in Austria but he did describe his canal project in a book he had published in London in 1900. Perhaps the best commentary on this particular canal is that of Baillie-Grohman himself who says in his book that the job was one that 'I can honestly recommend to those desirous of committing suicide in a gentlemanly manner.'[6]

Fort Frances Lock

Although there were other proposals for small canal links in British Columbia, none seems to have reached the stage of construction and so we must travel back in imagination to eastern Canada. This journey of almost two thousand miles was made in canoes by the pioneers, using the early waterways discussed briefly on p 2. One of the most difficult sections of this long route was that between Winnipeg and Lake Superior, through the area dominated by the Lake of the Woods. When steamship services on Lake Superior had begun to speed up and improve journeys westwards as far as the Lakehead, much attention was given to possible improvements of the long and tiresome journey from Prince Arthur's Landing (later Port Arthur and now part of Thunder Bay) to Lake Winnipeg and the Red River Settlement. A careful survey of the entire route was made in 1857 by Simon J. Dawson, a Scottish civil engineer. He suggested roads for the eastern and western ends, improvements for the portage roads, and the use of small steamboats on the larger lakes.

Early in the winter of 1869 he was given instructions to start work on what was to become known as the 'Dawson

Road'. The dispatch of an army expedition to quell the Red River uprising of the following year put the first part of the uncompleted road to good use in 1870 and focussed attention on the possibilities presented by the combined road and water route. Railway building was in the air, a start having been made by the Government of Canada, with direct labour, at building the transcontinental line that was eventually to be finished as the Canadian Pacific Railway. The mid-seventies in Canada were years of economic difficulties and so work was pushed ahead on the Dawson Road and it was well used, despite its limitations, for the few years remaining until the CPR was completed.

The building of the Canadian Pacific has so caught the public imagination, within Canada and far beyond, that this earlier attempt to provide improved transportation to the West is a chapter of Canadian history but little appreciated even by Canadians. It is, therefore, not too surprising that the inclusion in the Dawson Road of a navigation lock is a part of the overall story almost unknown. But a lock was proposed and almost $300,000 spent upon it. The location was at the outlet from Rainy Lake into the Rainy River, a strategic spot now occupied by the town of Fort Frances (named after Frances Simpson, wife of Sir George Simpson of the Hudson's Bay Company which maintained a post at this location). The corresponding US town is known as International Falls, indicative of the impediment to navigation provided here by the spectacular falls, the international boundary being formed by the Rainy River for about 30 miles until it discharges into the Lake of the Woods.

A modern bridge connects the two towns. As one crosses over this the excavation for the canal may be seen beneath, on the Canadian side of the Rainy River. It is about 800 ft long, in solid rock, and is now used as an overflow channel from the adjacent paper mill. A lock, 200 ft long by 38 ft wide with 5 ft 6 in over the sill, was to be constructed in this channel, work on which started in 1876. Total expenditure was $288,278.51, this modest sum being explained when we note that all work was stopped early in 1879, before the lock proper had been installed. This strange interruption can be explained only in political terms. The canal works were started under the Liberal administration headed by Alexander Mackenzie. On 9 October 1878 this Government was defeated and the Conservatives, under Sir John A. MacDonald, were returned to power, pledged to complete the Canadian Pacific Railway, and with no 'amphibious' connections such as Mackenzie had approved as an economy measure.

The unfinished canal has therefore stood for almost a century as a reminder of all the hopes then entertained for the water route through this lovely area. The paper company was permitted to build its dam, adjacent to the canal, in 1905 and it is from this date that the canal has been used as a waste channel. Many proposals have been made for completing the canal, even to the extent of establishing (1911) the Western Canal Company with this objective. Steamers sailed regularly between Kenora, on the Lake of the Woods, and Fort Frances until 1914 even though this meant breasting the Long Sault Rapids on the Rainy River, for the improvement of which the canal company proposed another small lock. With these improvements, a through water route from the east end of Rainy Lake to beyond Kenora would have been provided, a distance of about 200 miles. Although not yet built, this lock may yet prove to be one of Canada's modern canal works since the steady increase in the volume of pleasure craft in this wonderful lake area would appear to present warrant enough for the completion of a canal work started a century ago. And the unveiling of an historic plaque at Fort Frances, in September 1968, indicates clearly the local interest that still exists in this most isolated of all Canada's canals.[7]

Canals on Lake Ontario

Although Fort Frances is now in the far west of the province of Ontario, in the great days of canal building, and of canal fever, its location was still in the far west of all that there then was of developed Canada. Ontario was still Upper Canada until 1841; thereafter, and until Confederation in 1867, it was Canada West. Active settlement started around the shores of Lake Ontario and so it is not surprising to find a number of small canal projects carried out adjacent to the lake. The harbours of Toronto and Hamilton are each protected by interesting geological formations of accumulated sand. It was natural that channels should have been dredged through appropriate sections of these natural barriers to give access to the sheltered waters of the harbours. Strictly speaking, these openings were not really canals, although that into Burlington Bay—the harbour of Hamilton—was actually called the Burlington Bay Canal. It was 13 ft deep and 120 ft wide originally, although of course much enlarged since those early days. It was started in the 1830s and much improved after the Union of 1841, a sum of over $300,000 having been spent on it by 1852. Since Confederation it has been regarded as an integral part of Hamilton harbour.[8]

Such was the enthusiasm for canals in those early days, however, that a Mr Desjardins obtained a private Act of Incorporation in January 1826 for the construction of a canal, again without locks, from the inner end of Burlington Bay for 3.68 miles as far as the (then) small town of Dundas in order to give it access to Lake Ontario. The £ 10,000 required as formal capital was subscribed in the neighbourhood but a grant of $68,000 was made by the Legislature, bearing interest at 6 per cent, to assist with construction, the total cost coming to $98,684. It provided a depth of water of 9 ft and was 33 ft wide. Opened officially on 16 August 1837, it was never a paying proposition, the loan never being repaid. The coming of the railways sealed its doom but it was still carrying some freight up to the 1850s. Prior to that the (new) Great Western Railway found that the only way their line could cross the canal was by an embankment near its entrance.[9] For a time, therefore, the canal was closed but the railway company naturally had to provide a new opening, which they did by means of a fairly high level bridge. This achieved dubious fame as the scene of the second worst accident in the history of Canadian railways, a small train falling off the bridge on 12 March 1857 onto the ice then covering the canal. It is difficult today to imagine the scenes of horror that followed the accident, since the surroundings of the old canal now form a peaceful bird sanctuary.

At the other end of Lake Ontario, there may still be seen the remains of yet another comparatively unknown canal venture, this one actuated rather than defeated by the advent of the railways. The key location of Kingston at the eastern end of Lake Ontario was naturally mentioned in connection with the southern end of the Rideau Canal. Facing Kingston in the broad waters of the River St Lawrence as it leaves the lake is Wolfe Island, a narrow island about 14 miles long, separating the river into two main sections about three miles and one mile wide respectively. Some sort of ferry service had operated from the earliest days of settlement between Kingston and Wolfe Island and then from the south side of Wolfe Island to Cape Vincent on the American shore. The importance of this service was increased in the 1830s when Kingston was made the official customs 'Port of Entry' from and 'Port of Exit' to Cape Vincent for all the Canadian settlements between Cornwall and Port Hope (about 200 miles). The ferry became even more important when the railway reached Cape Vincent. Interested citizens conceived the idea of excavating a canal across Wolfe Island, in order to avoid the nuisance of transhipment. They had a survey made, the route being 2½ miles long, naturally with no locks, and prepared a petition that was addressed to the Governor, Sir John Colborne, on 25 February 1834. He commended it to the Legislature but, probably because this was near the time of the troubles (1837), nothing was done.

In 1846 the matter was revived and an Act passed incorporating the Wolfe Island, Kingston and Toronto Railway Company. Its name indicates the hopes behind it but the Act lapsed. Another Act was passed, however, incorporating the Wolfe Island Railway and Canal Company and this company did build the Wolfe Island Canal; it was finished in 1857. To begin with it had a depth of 6 ft, later increased to 7 ft. It was in regular use for the Kingston to Cape Vincent Ferry service for over 20 years but the service was given up before the end of the century and the canal allowed to fill up with weeds. There have been several suggestions, down the years, that it should be cleaned out and dredged again but nothing has been done and so the old canal remains as a strange weed-filled depression across this lovely island, not often even recognised for what it was. The company that constructed it originally has a special interest for Canadians since the great Canadian prime minister, Sir John A. Macdonald, when a young lawyer in Kingston, was associated with this canal venture.[10]

The Newmarket Canal

Railways, or rather railway freight rates, were responsible for what is perhaps the most remarkable of all these minor Canadian canals since it was constructed in the twentieth century. Railways were then well developed throughout eastern Canada, most of southern Ontario being served by the Grand Trunk Railway. This line gave good service but its 'public relations' (as they would be called today) were never satisfactory, if only because the line was managed from London, England. In 1904 the Grand Trunk raised its freight rates and there were loud public outcries. At the small town of Newmarket, located about 25 miles due north of Toronto, the protest took the form of a public meeting held at the town hall on Saturday afternoon, 10 September 1904. The Mayor spoke strongly and suggested that freight rates could be kept down if only the town had a water connection with the Trent Canal. The map on page 86 shows that a small river (the Holland) drains into Lake Simcoe at its southern end. Holland Landing is an old settlement with a name that shows how important a point it was at the northern end of the old portage road up from Lake Ontario. The Holland River was navigable from 'The Landing' into the lake, so much so that steamers had not only sailed to and from its wharf but the first steamer to sail on Lake Simcoe (the *Simcoe*) was built there in 1832. The idea of a branch canal to Newmarket therefore involved dredging the 9 miles up to Holland Landing from the lake, and then constructing 3 locks, with more dredging, in the 4 miles between The Landing and Newmarket. It had been discussed for many years and so the idea of building it was enthusiastically welcomed, not least by Sir William Mullock, a local member of Parliament and a member of the Government of the day.[11]

A large deputation went to interview the prime minister of Canada and members of his cabinet on 21 February 1905 and was favourably received. And the canal was actually built, though never used even though the channel had been dredged and all concrete work completed. The three concrete dams and lock structures stand forlorn and isolated, all but forgotten apart from that furthest upstream, where the retained water is the central feature of a local conservation area and nature sanctuary of real beauty.

Locally, the abandoned canal is still known as 'Aylesworth's Ditch', even though the prominent Canadian whose name is so used never had anything to do with the canal as far as can be ascertained. Canadians have unusual ways of memorialising some of their leading political figures. How anyone imagined that such a venture could possibly be commercially successful, especially since it could provide low freight rates only in the summer period of navigation, is today beyond belief.

The Newmarket Canal, however, is far more than just a reminder of one of the last outbreaks of canal fever in Canada. It is the only tangible evidence of all the hopes that persisted for over a century for a canal connection between Lake Ontario at, or near, Toronto and Lake Simcoe with its access to Georgian Bay, the Great Lakes and the West. As early as 1791 one of the partners in the North West Company (the great partnership of fur traders out of Montreal which eventually amalgamated with the Hudson's Bay Company) examined this route on foot. Eventually the company built a portage road, the beginnings of the famous Yonge Street of Toronto that is today that city's main south-north highway. The usefulness of the portage road, in eliminating the great haul over the Niagara escarpment until the coming of the first Welland Canal, naturally led to thoughts of what a canal could do as a supplement to the road. As early as 1800 a proposed canal was surveyed, to run between the Holland River and the Rouge River, a few miles to the east of Toronto. Proposal followed proposal, most of them favouring a canal connecting the Holland River with the Humber River which runs into Lake Ontario just to the west of Toronto Harbour. Meetings were held; deputations interviewed governments; reports were made; estimates were prepared, one in the 1850s suggesting a total cost of $22 million.

Nothing was done, however, apart from the building of the little Newmarket Canal which can now be seen to be a dubious and minor offspring from this century of discussion and promotion. The proponents of the Welland Canal, and so of the St Lawrence route to the Lakes, were very powerful as the results of their efforts were to show but there was also in the background, and throughout the century of argument, yet another proposal for circumventing Niagara Falls and one which presented advantages far outweighing all that could be mustered in support of the Toronto & Huron Canal, as the abortive suggestion of the canal to Lake Simcoe was generally known. This was the canalisation of the Ottawa, Mattawa and French Rivers—the Ottawa Waterway—which came within an ace of being constructed. To this greatest of the 'ifs' in Canadian history, we must now turn.

CHAPTER 7

The Great Dream
The Georgian Bay Ship Canal

This century-long attention given to developing a short route
for small vessels from Lake Huron to Lake Ontario by way
of Lake Simcoe indicates clearly the importance of water
routes to the West throughout the nineteenth century. From
the time of Champlain's first journey up the full length of the
Ottawa Waterway in 1615 until the coming of the railways,
the Ottawa route was the Gateway to the West. A glance at
the map on p 118 will show the significance of the water
highway that it provided; it is on almost a straight line from
the east end of Lake Superior all the way to Montreal. The
Ottawa River itself was used from Montreal as far as the
junction with the small Mattawa River, coming in as a tributary from the west. By going up the Mattawa, travellers
reached a height of land at Trout Lake only four miles from
Lake Nipissing, which drains into Lake Huron by way of the
French River—another result of the glacial geology of this
region. Portaging from Trout Lake into the tiny Vase River
and so into Lake Nipissing, those using this wonderful route
then followed the south shore of Lake Nipissing until it leads
into the French River, down which they proceeded into
Georgian Bay and so into Lake Huron, with the connections
it gives to Lakes Michigan and Superior. Up this waterway
came almost all the explorers of the west of Canada, of the
mid-west of the United States, of the Mississippi and Missouri Rivers, the discoverer of the Mackenzie River and
even of the Arctic coast, and the first white man to cross the

117

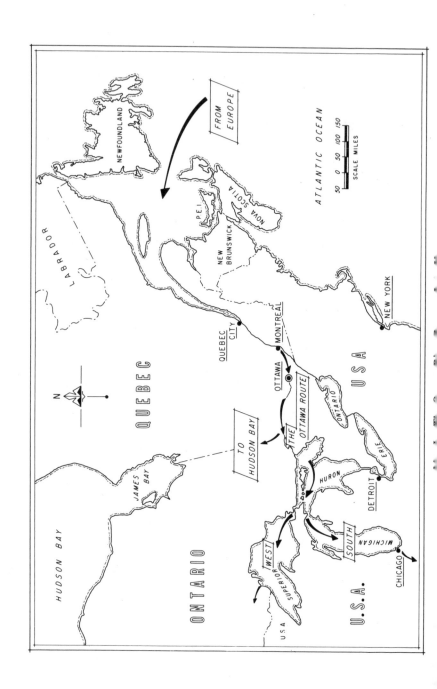

North American continent, Alexander Mackenzie.

This was the route to the West developed so remarkably by the North West Company in the building-up of their trade in furs, its use reaching a peak about the turn into the nineteenth century. Brigades then regularly left Montreal in the spring in their canoes with trading goods for fur-trading posts as far away as the valley of the Mackenzie River, well over two thousand miles away. Even after the amalgamation of the North West Company with the Hudson's Bay Company in 1821, it was still used for another half century by some of the fur traders for their special journeys. One such journey was the annual inspection trip made by the Governor, Sir George Simpson, from his home at Lachine. It was but natural, therefore, that when canals had become recognised as a valuable type of river improvement, thought should be given to the canalisation of the Ottawa route. The many portages around its majestic but difficult rapids and glorious waterfalls made the transit of the Ottawa Waterway an arduous journey, taking usually no less than two weeks. Replacement of these portages by dams and canal locks must have been the dream of many a *voyageur* as he laboured under the weight of his two 80 lb packages on the more difficult portages. Canalisation was suggested as early as 1827. For almost one hundred years thereafter, the Ottawa Canal—eventually designated the Georgian Bay Ship Canal in the early years of the twentieth century—was the dream of many Canadians, and especially of those up and down the valley of the Ottawa, and even of British financiers and builders who were attracted by its challenge and its potential.

The Ottawa provides a water route between Lake Huron and Montreal that is 300 miles shorter than the now better known route by way of Lake Erie, Niagara Falls, Lake Ontario and the St Lawrence. In view of this, it is not too surprising to find that the St Lawrence route was discovered almost simultaneously in 1669 by La Salle making his way slowly up the rapids on the St Lawrence, and by Joliet who was told about it by Indians with whom he travelled up the Ottawa route with instructions from the great French Intendant Jean Talon to find, if possible, an easier way of getting down to Montreal from the upper Great Lakes than by way of the Ottawa. The St Lawrence route is indeed the 'easier' route, with fewer rapids, much of the fall concentrated in the great drop at Niagara, and with clear sailing on the Lakes. The Ottawa route necessitates a rise of 659 ft to the summit level and then a subsequent drop of 99 ft from Lake Nipissing to Georgian Bay, this drop being wasted effort in a way especially when considered as part of a canalisation scheme. But balanced against this is that saving of 300 miles and, of indefinite but very real significance, the fact that the Ottawa route would provide a seaway to the Lakes that was all in Canadian territory. Despite this, it has been the St Lawrence route that has been developed as the Seaway. Prominent among the reasons for this turn of fortune's wheel is the fact that settlement, and corresponding development, came first to the shores of Lakes Ontario and Erie, as Upper Canada was first established and then became Canada West, eventually Ontario. The Ottawa valley then, as now to some extent, has been a backwater as far as provincial administration has been concerned. John Graves Simcoe, for example, the first Governor of Upper Canada, despite his admirable and remarkable activity, never even visited the valley.

The story of the vicissitudes of the Ottawa River canalisation proposal is, naturally and inevitably, long and complex. It aroused bitter feelings at times, as between the proponents of the St Lawrence route and those favouring the route up the valley. It gave Canada a taste of really vituperative pamphleteering. Royal Commissions galore investigated it and reported upon it or, on occasion, dodged the issue of making any recommendation, so strong were public feelings. Throughout the century, defence authorities remained amongst the most ardent supporters of the Ottawa route since it was all-Canadian and avoided the difficulties that the use of the international section of the St Lawrence always created in the minds of those responsible for defence planning. This meant British defence authorities until well on into the century since defence was the last governmental responsibility to be fully accepted by Canada from Great Britain. The last resident British troops left Canada in 1906 but the naval dockyards at Halifax and Esquimalt (on the Pacific coast) remained under the Royal Navy until 1912. It is interesting, therefore, to find a succession of reports by officers of the British Army and the Royal Navy, sometimes reporting jointly, on measures for the defence of Canada (always against the United States!) in almost every one of which there is a strong plea for the completion of the canalisation of the Ottawa River. To begin with, this meant more canals similar to the small Ottawa River Canals but, after a Seaway capable of conveying ocean ships had gained serious attention, it was for a Seaway up the Ottawa that defence authorities pleaded so forcefully. Some readers will probably be incredulous when I note that the last report of this kind, now arguing for the building of the Georgian Bay Ship Canal, was submitted as late as 1912.

The first real survey of the entire route of the proposed canal was made in 1856 by Walter Shanley, an eminent civil engineer, as an extension of the inspection of the Ottawa River Canals which he was asked to make prior to their takeover by the Government of the Province of Canada. He estimated the total cost, after a personal inspection of the entire route and some survey work by his assistants, as $24 million. This high figure must have caused consternation in official circles since another survey was made by another civil engineer, T.C. Clarke, in 1859. His estimate for the complete project was only $12 million. The difference between the two figures is readily explained by the overall plans of the two men, Shanley relying on excavation whereas Clarke reckoned on building impounding dams in association with his locks. It can be readily imagined how these two estimates were bandied about in ensuing discussions, with never a word to explain the difference! The two figures were even used by one of the many Royal Commissions on Canals to explain why they could make no comment on the advisability of going ahead with the Ottawa River canalisation, one of the most remarkable excuses for 'dodging a hot potato' that one is likely to come across. Enthusiasm ebbed and flowed during the latter part of the century, a major revival of interest taking place in the final decade. A Committee of the Senate of Canada was established in 1896 specifically to report on the Ottawa River project. It reported favourably in less than three months from the date of its appointment, official bodies working fast in those days. The most remarkable feature of its work, however, was not its report but the fact that one of the witnesses it heard was a Mr Meldrum, on the staff of S. Pearson and Company, the great contractors of London, England, who stated that his company was ready to undertake the building of the proposed canal if suitable financial guarantees could be assured.

This put a different complexion on what had been up to then a rather theoretical proposal. It placed the Montreal, Ottawa & Georgian Bay Canal Company in public favour. The company had been chartered in 1894 and for the next 30 years it was to be at the centre of the tangled negotiations that continued unabated until its charter was allowed to lapse. As the twentieth century opened, interest in the canalisation of the Ottawa approached its peak. Another Royal Commission was appointed but it managed to avoid making any commitment about the controversial canal by referring to the fact that the Government of Canada had arranged for an entirely new survey and study of the proposed canal. They had indeed, the engineering staff of the Department of Public Works having been instructed in 1904 by their minister, with authority from Parliament, to expend up to $250,000 on an entirely new survey for a ship canal from Montreal to Lake Huron by way of the Ottawa, Mattawa and French Rivers. This great task was placed under the direction of a notable civil engineer, Arthur St Laurent, who was capably assisted by two remarkable assistants whom I mention since, when I was starting my own enginering work in Canada, I had the happy privilege of coming to know one of these men. I heard from S.J. Chapleau in person, sitting at his desk, something of the quite remarkable work that was done in preparing the report on this major undertaking. It finally cost about twice the original Parliamentary authorisation but never was public money better spent.[1]

The final *Report upon the Survey for the Georgian Bay Ship Canal, with Plans and Estimates of Cost, 1908*, presented to the Parliament of Canada in its 1909 session, is one of the greatest of all Canadian engineering documents. Although published in the dull-looking form of a Sessional Paper, this 600-page volume presents a clear and vivid picture of the ship canal that could be built, its appendices answering every possible question that would normally be raised in discussion of its proposals. The first conclusion presented in the second paragraph of the Letter of Transmittal says: 'That a 22-foot waterway for the largest lake boats (600 ft x 60 ft x 20 ft draft) can be established for one hundred million dollars ($100,000,000) in ten years, and that the annual maintenance will be approximately $900,000, including the operation of storage reservoirs for the better distribution of the flood waters of the Ottawa River.' No beating about the bush! This was the considered and unanimous conclusion of the very able group of engineers responsible for this great study, the extent of their activities being perhaps indicated by the fact that no less than 2,990 test borings were taken at the sites of proposed dams and locks all along the waterway. The size of locks differs only slightly from those today in use on the St Lawrence Seaway, yet they were proposed over 60 years ago, clear indication of the forward thinking of Mr St Laurent and his colleagues. The Report was naturally widely hailed by all supporters of the Ottawa route; equally naturally, it was attacked by those who favoured the St Lawrence route. It was so clear a description of the Seaway that could be built up the Ottawa that the Government had to indicate some decision without too much delay.[2]

Sir Wilfrid Laurier was the prime minister of the day. The times had been good economically and so his Government had embarked on some heavy expenditures of public money, notably for railways. Problems were arising, however, and not only in connection with railways. Sir Wilfrid often mentioned the Ottawa scheme in his speeches—he could do no other!—and clearly indicated that, if his Government were to be re-elected in the forthcoming general election, the Georgian Bay Ship Canal would be built. Politics being what they are, in Canada at least, the election of 1911 was fought on an emotionally charged issue of possible economic reciprocity with the United States; it is still often referred to as the Reciprocity Election. The ship canal was not a major issue, although always in the background, but it went down to defeat when the Liberal Party were defeated by the Conservatives under Sir Robert Borden. This turn in political fortunes led to an entirely new situation, especially as the Borden Government was faced not only with internal financial problems brought on by the lavish expenditures on railways but also by the approach of the outbreak of World War I. Delegations continued to come to Ottawa to plead the cause of the Georgian Bay Canal but somewhat naturally they met with polite although entirely non-committal replies. The demands of war soon dominated everything and the Ottawa Seaway had to stand aside.

Once the war was over, agitation started again, notably by officers of the company which still had its charter. The whole situation was now complicated, however, by an entirely new factor—the value of the water power that could now be generated so easily by the very falls that the canal locks were to circumvent. Public power development was already well recognised in Ontario whereas private companies were responsible for the first major water power projects in Quebec. Both provinces considered that the water power available on the Ottawa River was theirs to develop and not under the jurisdiction of the Government of Canada. The consequent legal tangles must be left to the reader's imagination but the grand finale of the Montreal, Ottawa & Georgian Bay Canal Company came in 1927, after control of the company had passed from British into Canadian hands. The battle of the public versus private power generation question was fought again in the House of Commons, and before one of its important committees, as the renewal of the company's charter was debated. The debate strayed a long way from merely the charter renewal; four whole days were taken up with one of the most bitter debates that the Parliament of Canada had witnessed for a long time. Eventually, the private bill to renew the charter did not pass and so the charter lapsed. But the debate continued as did also the legal arguments. Eventually, the two provinces were granted the necessary permission for developing the Ottawa. In subsequent years they worked out a statesmanlike agreement whereby each province was to develop a certain number of the main power sites without reference to the interprovincial boundary which runs down the centre of the Ottawa. This procedure has already been followed for four of the main sites, only the first major plant (at Chats Falls, about 30 miles upstream of Ottawa) having been built with the power house exactly straddling the boundary, power generated in each section being for use by the respective province. And only one of the plants (that at Carillon, p69) incorporates a navigation lock.

But the great dream of the Georgian Bay Ship Canal refuses to die. Organisations, especially in the Ottawa Valley, continue to advocate its reconsideration; the existence of the great man-made lakes now formed by the great dams that impound the river for power generation serve to remind all who see them of what the Ottawa River might have looked like had the proposal of 1908 been carried out. Had it been built, the need for the St Lawrence Seaway would not have been so pressing. The Ottawa Valley would have been developed far more than has so far been the case. And the fact that the Seaway would have been entirely in Canadian territory would have had a profound effect upon the economic development of all of north-eastern North America. But all this is dependent upon the great 'if' and the answer, as the world knows, went the other way. It remains to ask whether anything at all in the way of canalisation was carried out above the Chaudière Falls at Ottawa even though the Georgian Bay Canal still remains a dream for the future. Two small projects were started; one was completed. They must not be passed by.

The next great falls on the Ottawa River above the capital city were the Chats Falls at the head of a broad expanse of the river known as Lac Deschenes. They were a magnificent sight, the river falling about 50 ft over a mass of rock outcrops and small islands. So concentrated was the drop at the falls that it was easy to imagine a canal around one end of the falls which, with suitable locks, would have enabled vessels to pass between Lac Deschenes and Chats Lake above the falls. This possibility was not lost on early visionaries in the valley who managed to persuade local political figures of note that the Government of Canada should set about constructing such a canal as a part of the eventual canalisation of the whole river. Work was therefore started in 1854 on the necessary excavation and, to balance political considerations on both sides of the river, a quarry for building stone for the locks was opened up on the south bank. The only real problem was the local geology, since the canal was sited on some the hardest igneous rock of Precambrian age that can be imagined. With the tools then available, progress was very slow; complaints from the contractor mounted with each passing month. Eventually the work was stopped but not before almost $500,000 had been spent on this abortive effort. The excavation that was done can be seen today in what is still an isolated spot on the north bank; the quarry, on private property, can also be distinguished. Transfer around Chats Falls, from steamers on the two lakes, had to continue until the coming of the railways by means of a 3 mile horse-drawn tramway, yet another unique feature of the Ottawa valley.

Equally remarkable was the political history and eventual building of two locks in tandem on the Culbute Channel of the Ottawa, about 60 miles upstream from Chats Falls. The project was intended to assist steamboat navigation on this upper section of the Ottawa but it was so political in nature that its story is a tangled tale indeed. Two locks were built in the 1870s, all of timber as was the necessary control dam at the adjacent rapids, the locks being opened for use at the start of the 1877 navigation season. They were built 200 ft by 45 ft with 6 ft of water over the sills. Once again, railways rendered useless the expenditure here of about $250,000 since the first rail line reached the nearby town of Pembroke in 1879, all passenger steamship services being withdrawn the following year. The Culbute locks continued in desultory use for a few years but by 1888 their abandonment (the word was used officially) was being recommended. They were actually abandoned in 1889 and left to the care of nature, their sad story being completed in 1912 when much of the remaining exposed timber work was destroyed by fire.[3] Today, the remains of the lock and dam lie deep in the bush of this still remote part of the Ottawa valley, timbers under water being still in good shape. To stand by the side of this practically unknown and derelict canal project, as I have done on a day when heavy rain dripping through the trees added its own poignancy to the scene, is a moving experience. One thinks readily of all the hopes entertained for this lock system and of the grander hopes for the Georgian Bay Ship Canal, which would have so transformed the whole of the Ottawa valley, but then one remembers that the Seaway was built far to the south of Culbute, on the great St Lawrence River, to the canalisation of which we must now give our undivided attention.

PART II

CHAPTER 8

The St Lawrence System

The St Lawrence River-Great Lakes system is one of the most wonderful major river basins of the world. The Gulf of St Lawrence and that portion of the river downstream of Montreal are themselves remarkable for making the metropolis of Canada a leading ocean port even though it is about 1,000 miles from the open sea. We are concerned, however, with that portion of the St Lawrence River upstream of Montreal, with its connecting Great Lakes and the rivers that make these connections, all really a part of the St Lawrence even though they may bear other names.

The total drainage area of the St Lawrence-Great Lakes system is 678,000 square miles, almost twelve times the size of England and Wales. Somewhat less than one half of this vast area lies upstream of the exit from Lake Ontario. Of this still great area, no less than 95,000 square miles or almost one third represent the area of open water on the Great Lakes, this area being greater than that of England, Wales and Scotland combined. From Montreal to Duluth at the western end of Lake Superior is a distance of over 1,300 miles giving a total length for the river above Montreal, allowing for its tributaries into Lake Superior, of about 1,500 miles. The balancing effect of the Great Lakes gives to the St Lawrence another of its unique features, the ratio of its high water flow to its annual average flow being only two; the significance of this very uniform flow cannot be overestimated.

Vast though these statistics show the St Lawrence system

to be, it ranks only about twenty-second amongst the main rivers of the world. It is not even the largest river of Canada, although few Canadians realise this. The Mackenzie River, rising in the Rocky Mountains and flowing generally northwards to enter the Beaufort Sea on the Arctic coast through a great delta, now the site of the new town of Inuvik, is longer and in other respects a larger river than the St Lawrence. The special significance of the St Lawrence system, however, is that it penetrates to the heartland of industrial Canada and the United States. This continental core, industrial in strategic areas and agricultural in areas away from the Lakes, can be said to have an area of well over one million square miles, now with a population of over 60 million people. Yet another special feature of the St Lawrence is its international character, the boundary between Canada and the United States lying in the connecting rivers and running through the Lakes between 75°W and about 90°W, for a total distance of well over 1,000 miles. This renders all the Great Lakes, with the exception of Lake Michigan (wholly within US territory), of joint concern to both Canada and the United States. It was this joint concern that led to the establishment of the IJC (see p 18) whose major responsibility is naturally surveillance of the St Lawrence-Great Lakes system.

It may be helpful by way of introduction to mention the main cities that lie on the Great Lakes, their location on this great waterway system having certainly been responsible, at least in part, for their growth and their importance today. Montreal is our starting point, now a metropolitan area with well over two million people living on the island of Montreal. Near the start of the international section lie the small cities of Massena (US) and Cornwall (Canada), each an important industrial centre. On Lake Ontario, Oswego and Rochester are the major US ports, with Syracuse only 25 miles from the lake; on the Canadian side, Kingston, an historic city, lies at the exit from the lake, with Toronto and Hamilton, now great industrial centres, at the western end. Buffalo stands at the exit of Lake Erie into the Niagara River, Fort Erie facing it in Canada, and on the shores of this lake are Cleveland and Toledo, each a major commercial centre. Detroit, with its vast automobile and other industrial plants, stands on the Detroit River connecting Lake Huron with Lake Erie, Windsor in Canada facing it (but to the south!). Sarnia in Canada and Port Huron stand at the exit from Lake Huron but on this lake there are no major cities although many smaller fine communities. Sault Ste Marie, at the exit from Lake Superior, is the name borne by both Canadian and US cities. At the western end of this greatest of the lakes stands the newly united city of Thunder Bay in Canada (formerly Fort William and Port Arthur), and Duluth and Superior even farther to the west, in Minnesota and Wisconsin respectively, at the extreme head of the lake. On Lake Michigan, Chicago calls for special mention since by using the locks on its so-called Drainage Canal, vessels can gain access to the Mississippi River system and so to the Gulf of Mexico. Milwaukee is another important centre, the shores of Lake Michigan being now well developed and in places highly industrialised.

These great urban centres will call for mention in the pages that follow which is warrant for listing them here, even though most of them are developments of the last 150 years. For more than one hundred years, however, they have had access by water to the Atlantic Ocean by way of the St Lawrence. Vessels coming into the lakes from the sea were at first naturally limited in size to relatively small craft but for much of this century there have been regular sailings from European ports into the Great Lakes of specially designed steamships, and later diesel vessels, again limited in size by the locks on the St Lawrence but large enough to provide a reliable trans-Atlantic freight service. With the opening in 1959 of what is now called the St Lawrence Seaway, the size of ocean vessel that could sail up to the Great Lakes was dramatically increased so that vessels of leading steamship lines can now be seen over 2,000 miles from the open sea.

These same limitations of size have not always applied to vessels sailing between Lake Superior and Lakes Huron and Erie nor, since 1932, into Lake Ontario by means of the latest Welland Canal. A great fleet of immense vessels, designed specifically for service only on the Great Lakes, has been for many years another distinguishing feature of this part of the waterway. Its activity is indicated by the fact that, despite an open season of navigation restricted by winter conditions for three months or more, the locks at Sault Ste Marie now pass each year a tonnage greater than that handled in twelve month seasons by the Panama and Kiel Canals combined. And in the navigation season of 1973, the new Seaway locks on the St Lawrence passed well over 50 million tons of freight.

The St Lawrence system is, therefore, one of the really great canals of the world. It will be convenient to describe it section by section since the locks and dredged channels are separated by long stretches of clear sailing through the lakes. We will make an imaginary journey through the new Seaway and up to Thunder Bay, in conclusion, but before we study the canalisation of the St Lawrence, it will be well to take a brief glance at the natural state of this waterway before it came into wide use by white men. This period of use is a short span in the pattern of human history, not yet four hundred years, and yet discovery of the St Lawrence goes back to the earliest days of North American settlement.

It was on 2 October 1535, forty years after Columbus, that Jacques Cartier, during his second visit to Canada, first landed at what he called Hochelaga, to be greeted by a thousand Indians who made him welcome. This was the site of Montreal, not to be founded as a permanent settlement until 1635, but visited and recorded by this intrepid master-mariner from St Malo. He was stopped in his small craft at what is now the island of Montreal by the great rapids in which the St Lawrence drops about 50 ft as it flows down from a wide smooth expanse now known as Lake St Louis. Cartier must have seen these rapids, just as they are today, and he was also told about the great river to the west, this being the Ottawa and not the St Lawrence. It would not be until 1613 that a white man would record what lies to the west of the great rapids. This was Samuel de Champlain, the fine man who was the real founder of the Canada of today. When he made his first visit to Montreal, the village of Hochelaga had disappeared. He shot the Lachine Rapids in an Indian canoe, as a mark of special favour by his Indian hosts.

The head of Lake St Louis is about 12 miles upstream from the head of the Lachine Rapids. Here the Ottawa joins the St Lawrence (see map on p 52) through two turbulent channels on either side of Ile Perrot. The St Lawrence comes in to the lake just to the south of the Vaudreuil Channel of the Ottawa, coming down a series of turbulent rapids that extended intermittently for 16 miles down from the next wide expanse of the river, known today as Lake St Francis. Although in earlier days there was no international boundary anywhere near the river, the present border line between Canada and the United States comes to the river near the west end of this lake so that the stretch of river from here to Lake Ontario is that now known as the International Rapids section. Rapids there were indeed, dominated by the Long Sault just to the west of Cornwall, a magnificent turbulent stretch of really 'white water', untamed until the recent completion of the great power station and Seaway locks that have transformed this entire section of river into another peaceful lake. The total fall from Lake Ontario was about 90 ft, of which 50 ft were in the Long Sault. A sail of 180 miles takes us to the west end of Lake Ontario where the St Lawrence has its first name change, to the Niagara River. It flows smoothly out into the lake with little evidence of the great rapids just a mile or two upstream. The river flows swiftly for four miles below the world-famous 326-ft Falls. The Niagara escarpment, over which the river tumbles in the great Falls, dominates the landscape along the west end of Lake Ontario, extending as a distinctive physiographic feature almost up to Georgian Bay in Lake Huron.

Niagara Falls provided the greatest barrier of all to early water travel up the St Lawrence and Great Lakes route. Portages around it were therefore a contribution of the Indians to the convenience of the first European travellers. We shall see later how successive canals have circumvented this magnificent natural wonder. Once in Lake Erie, the way ahead was clear sailing, for about 230 miles almost due west across this shallow lake, then after a swing northwards along a winding channel for 70 miles up the Detroit River and into Lake St Clair (a small enlargement) and then up the St Clair River into Lake Huron. Another sail of over 200 miles up this lake would bring us, past fascinating islands, into the winding St Mary's River. Here a 20 ft drop out of Lake Superior was concentrated in the rapids named Sault Ste Marie by the early Jesuit priest-explorers. Once past these rapids, one was in Lake Superior, truly an inland sea, with a clear sail of almost 400 miles to its western end and deep water all the way (the bottom of Lake Superior being well below sea level).

This was the St Lawrence Waterway as nature had provided it through recent geological processes that are themselves a fascinating story. So recently was the waterway formed in its present state that the ground around the lakes, especially to the north of Lake Superior, is still rising very slowly as it recovers from the load of continental ice that it bore until about 11,000 years ago. So useful have the spillways used by the discharges from lakes dammed up by this ice proved to be that two of them have been utilised to convey water from rivers naturally flowing into Hudson Bay down to the north shore of Lake Superior. This Arctic water now helps to generate power in the great stations at Niagara and Cornwall-Massena.

To summarise, and remembering that all the lakes that call for mention are themselves parts of the course of the St Lawrence River on its way from beyond the head of Lake Superior to the sea, let me recapitulate the main features of the waterway. Going upstream from Montreal, the Lachine

Page 135. *Third Welland canal: a view in the harbour of Port Dalhousie in 1904, showing a steamer arriving from Toronto with passengers for the train to Niagara Falls. This trip by steamer across Lake Ontario and by train southwards and back was a popular excursion until the years of the second world war*

Page 136. (Above) *Fourth (modern) Welland canal; excavation in progress for the flight of twin locks up the Niagara escarpment (Nos 4, 5 and 6) as seen on 7 July 1916 just before all construction work was stopped because of the first world war; (below) Lachine canal: S S Fairlake, a typical 'lower laker' or 'fourteen footer', moves upstream from the harbour of Montreal in July 1932 en route to Lake Ontario. This canal is now disused because of the Seaway*

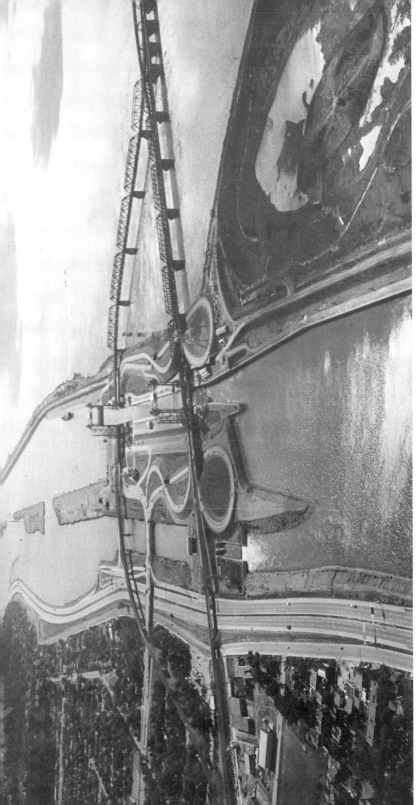

Page 137. Aerial view of the St Lambert locks along the St Lawrence Seaway, 1966

Page 138. (Above) *St Lawrence Seaway: typical 'graffiti' on the walkway of a berth downstream of one of the locks. Vessels tie up here to wait their turn to enter the lock;* (below) *Fourth Welland canal: an aerial view of a part of the new bypass channel opened in 1973. Here it is crossed by a main highway and important railway in a joint tunnel constructed before the canal was flooded*

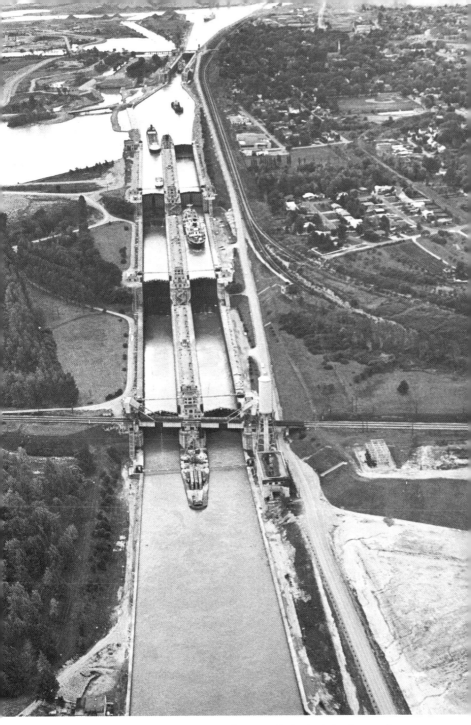

Page 139. *Fourth Welland canal: the famous 'flight locks', three twin locks making a gigantic marine staircase, near St Catharines and Thorold. They lift vessels up almost the full height of the Niagara escarpment*

Page 140. *Fourth Welland canal, 1966: the 'flight locks' seen at night*

Page 141. *Fourth Welland canal: the control room with its television monitors on all locks. Operators control the movements of all vessels in the canal by radio contacts*

Fig. 142. Aera aview of a site s. Site s Complex I. I. S. Complex Site. Marginal tower site of the roof of the Complex I. I. I.

Rapids had to be portaged in order to reach Lake St Louis. The St Lawrence entered the head of this lake through turbulent rapids, three groups of them constituting what came to be known as the Soulanges section. Lake St Francis followed and after a pleasant sail past small islands in the river, the Long Sault had to be passed near the present city of Cornwall. This was the great impediment on this, now the international, section of the river, the rapids beyond being minor in comparison, the river above them being at the level of Lake Ontario. From Lake Ontario the Niagara escarpment, with the great falls and rapids above and below, had to be breasted before Lake Erie was reached. Thence it was clear sailing into Lake Huron and Lake Michigan, with rapids on the St Mary's River to be passed before Lake Superior would be entered. These rapids, the fall at Niagara, and the rapids on the St Lawrence proper had all to be canalised before vessels could sail up from or down to Montreal and the sea.

Behind all the early Canadian canalisation projects was the spectre of lake traffic being diverted into the Erie Canal. This project in the United States extended from Albany on the Hudson River to Buffalo on Lake Erie and was officially opened on 26 October 1825, later having a short branch added linking it with Lake Ontario at Oswego (near Kingston). The building of the original Erie Canal is a fine part of the history of US canals. Its significance to Canada will be realised if the date of its opening be kept in mind as we consider together the rather halting start of its Canadian competitors.[1]

CHAPTER 9

The Lachine Canal

The Lachine Canal is the appropriate starting point for our study of the St Lawrence system not only geographically but also historically. It was in the year 1680 that Dollier de Casson, Superior of the Sulpicians in the tiny settlement of Montreal, first proposed the idea of digging a small canal to connect a small lake on the little St Pierre River with a bay on Lake St Louis. As we saw on p 2, the Ottawa River was the early gateway to the continent; it was being well used by Indians and French *voyageurs* by 1680, all of whom had either to negotiate the Lachine Rapids or portage their loads over the nine miles between Montreal and the lake. If they portaged their canoes also, they could use the small St Pierre River for the first part of the portage. The canal project involved a ditch (for it was little more than that) about one mile long, 12 ft wide and a minimum of 18 in deep. It was not until 1700 that a contract was signed for the work with Gedeon de Catalogne, a 'contractor' of Montreal. Unfortunately the contractor failed in 1701 when only 800 yd of the cutting still remained to be finished but the Sulpicians were unble to raise the necessary funds for its completion. Even the interest of the great King Louis XIV in 1717 failed to ensure its completion—but even though incomplete, the little channel was probably used by canoes at periods of high water. It would appear, therefore, to have some claim to being the first North American canal even though it was to be merely a channel without any locks.[1]

Sir Frederick Haldimand was the Governor of the new

144

British colony of Quebec, after the transfer of power, from 1778 to 1786. He has been criticized by some historians for his administration but he did effect improvements to the fledgling colony including, as we shall shortly see (p 152), the first small canals on the St Lawrence above Lake St Louis. These were built between 1779 and 1783. It seems entirely improbable that Haldimand would go to the trouble and expense of carrying out these works if he had not also provided an equivalent improvement at Lachine. The distance was almost 9 miles from the harbour of Montreal to the nearest point on Lake St Louis, the difference in water level being normally about 46 ft. Construction of a canal would not have been a difficult matter, therefore, especially since the land between, on which the portage road was located, is fairly flat. There seems warrant from some indirect reports that a new canal was built at Lachine at the same time as those to the west of Lake St Louis but it must have been a small affair, one report suggesting that the locks were only 16 ft wide. Unfortunately, no accurate records exist, this being one of the periods in Canadian history about which we would like to know much more but for which there are few records.

The 1783 canal may, therefore, be regarded as the first real Lachine Canal but the fact that it would have been designed to give passage only to the *bateaux* then in common use on the St Lawrence confirms the supposition that it was a small affair. With the advent of steamships the whole picture of water transportation changed radically. The first steamboat on the St Lawrence was the *Accommodation* of John Molson, launched in 1809 for service between Montreal and Quebec. Agitation for a new and better Lachine Canal can be traced back almost to this date. It was the War of 1812, again, that demonstrated clearly the need for a water connection from Montreal to Lachine other than the great rapids. Sir George Prevost, the new Governor, recommended the excavation of a real canal and an Act was passed in 1815 appropriating £25,000 for the building of such a work. With the coming of peace, however, action was again delayed with the result that in 1819 a joint stock company was organised by merchants of Montreal to build a canal with authority to raise capital to the amount of £ 150,000. This effort also failed but in 1821 the Government of Lower Canada got down to business, arranged to reimburse subscribers to the company and appointed Commissioners to undertake the building, the Imperial Government (as it was then called) making a grant of £ 10,000 on condition that all military stores be conveyed free through the canal when completed.

Work started on 17 July 1821 when the chairman of the Commission, Hon John Richardson, turned the first sod. The work was done by contract, under the direction of Thomas Burnett as engineer. Two of the three contracting firms went on to undertake more extensive works in the building of the Rideau Canal, on which their Lachine experience stood them in good stead. The works were completed in 1825 but it is believed that the canal was in use during the 1824 open season. It had 6 locks, dimensions being 100 ft by 20 ft with 5 ft of water over the sills with an additional guard or river lock. The canal itself was 28 ft wide at the bottom and 48 ft at water level, these dimensions being greater than those then generally in use in Great Britain. The canal was well used from the time of its opening. Indicative of the demands upon it is the fact that in 1839 it was opened for use as early as 11 April, while in 1852 (after it had been enlarged) it stayed open as late as 16 December. In 1849 it remained open for a period of 234 days, from 21 April until 10 December, dates which are remarkable for that period when no modern aids to winter navigation were available. The enlargement noted was loudly demanded because the canal had so quickly proved its utility.[2]

It was not until after the union of the two Canadas in 1841, and the consequent establishment of the Board of Public Works, that the claims for an improved canal were heeded. In 1843, however, the work of rehabilitation started but navigation through the canal could not be interfered with, so busy was the traffic; the new locks were therefore built 'in the dry' adjacent to the existing locks, the new channel being excavated in sections, this part of the work not being finished until 1848. Five locks replaced the original six, their width being 45 ft and length 200 ft. The canal was later widened so that two vessels could pass safely, further indication of its extensive use, this work being completed only in 1862. By this time, the St Lawrence route to the Great Lakes could be seen to be of vital importance to the future of Canadian commerce and so, at the time of the Confederation of the four original provinces, standardisation of the dimensions of all the locks on the St Lawrence was seen to be essential. More work therefore started on the Lachine Canal in 1863, four years before Confederation. Four new locks were built, again adjacent to the existing locks which were rebuilt and incorporated as twin locks; the fifth lock, the guard lock at the Lachine end, was built about 250 ft away from the 1842 lock and a new entrance into Lake St Louis was also constructed. This work had to be scheduled so that use of the canal would not be interfered with, a fact which explains why the new canal works were not completed until 1884. Some further work at individual locks was carried out subsequently but the canal as reconstructed at that time is that which served until 1959 when its place was taken by the St Lawrence Seaway. Its locks were 270 ft long by 45 ft wide with 17 ft available throughout the lower section of the canal, although locks 3, 4 and 5 in the upper part provided only 14 ft over sills. The canal was confined between masonry walls throughout its length so that its bottom width of 140 ft was only 10 ft less than its width at water level.[3]

The canal was operated and lighted electrically and even had its own small water power plant (near lock 4) which was maintained for emergency supply should the public supply from Hydro Quebec ever fail. It is crossed, as might be expected, by a number of notable bridges—vertical lift, swing and bascule. These carry main roads, the main lines of Canadian National Railways leading into Montreal's Central

Station, the Canadian Pacific Railway and branch railway lines. Two tunnels carry other roads under the canal while a third tunnel had to be provided for major water mains of the Montreal water supply system. Two major syphons also go beneath the canal, all these works, in addition to the fact that the canal is walled throughout, indicationg clearly the industrial complex through which the canal now runs. To some extent, this industrial development can be attributed to the canal itself, a major gas works (for example) obtaining all its coal supplies in earlier days by water delivery on the canal. Possibly the most interesting of the many industries that line the canal is the large steel fabricating plant of Dominion Bridge Company Ltd, from which loads too large to be handled in other ways could readily be shipped by water (see p 71). Since the canal was superseded by the Seaway, the responsible authorities have naturally wished to close it completely. The lower end, at Montreal Harbour, has indeed been filled in—alas!—but up to 1973 successive legal injunctions had managed to stave off the closure of the upper or western section into which vessels can still sail.

The Lachine Canal for over 130 years was, therefore, one of the most active of all Canadian canals and a vital part of the industrial geography of Montreal. Not only was it the entrance to the St Lawrence route up to the Great Lakes but it was also the first part of the journey from Montreal up the Ottawa River. Its entrance wharves in Lake St Louis, in the great days of travel by steamboat, used regularly to see the arrival of steamers from the Ottawa and also from the St Lawrence. Some of these fine vessels would disembark their more timid passengers at Lachine who could then complete their journey to Montreal by train. Those who stayed aboard, however, would be given the thrill of shooting the Lachine Rapids, in vessels which were a fair size, under the guidance of most skilled pilots, perhaps the best known of whom was Captain Joseph Edouard Ouellette, a direct descendant of the first white man to undertake this hazardous short trip after training by the almost mythical Indian, Big

John Canadian. This was an experience that one could enjoy (as I once did) until just before the second world war, an experience naturally never to be forgotten. After shooting the rapids the vessels would disembark their passengers in the harbour of Montreal, returning up the canal in the evening for the upstream return journey the next morning.

The St Lawrence Seaway has now so completely overshadowed the Lachine Canal that it must be difficult for young Canadians, let alone those who have not visited it, to appreciate what an important—indeed what a vital—link it constituted in the transportation system of Canada and in the industrial life of Montreal. Far from beautiful—how could it be, set amid a great industrial complex?—the little 9 mile long canal seemed always to be busy so long as navigation was possible. Not only did it carry all water traffic to and from the Great Lakes, and up and down the Ottawa, but its lower section was regularly used by ocean vessels, larger than could traverse its full length, bringing goods or loading outbound freight from the industrial wharves specially built along the first miles of the canal. It is probably inevitable that much of it will have to be filled in, as adjacent industrial plants make alternative arrangements for their transport needs, but it is greatly to be hoped that some parts of the canal, and some of its locks, will be preserved. Already, the famous Hudson's Bay House at the Lachine end of the canal, the home of Sir George Simpson for over thirty years, has been destroyed. The long service to Canada given by this oldest of all Canadian canals is warrant indeed for its partial preservation so that its service may be remembered as it takes its place as a part of Canadian industrial archeological heritage.

CHAPTER 10

The St Lawrence Canals

The St Lawrence River drops 187 ft in flowing from Lake Ontario down to Lake St Louis above Montreal. Of the total, 93 ft come between Lake Ontario and Lake St Francis, an enlargement of the river just east of the International Rapids section. The remaining 94 ft constitute the drop thence to Lake St Louis. Most of the first drop in level of the great river is now used for the generation of power in the international power house at Cornwall-Massena, built in association with the St Lawrence Seaway. The second part of the fall is largely used for a similar purpose in the Beauharnois power house, now owned by Hydro Quebec, located at the eastern end of a great canal which is used as a part of the Seaway. There have been great changes in recent years, therefore, in this 160 mile stretch of river, and especially since the building of the Seaway. It is our purpose in this chapter to see how the river was used before this, leaving an account of more recent developments for later chapters.

The bare figures of the drop in level conceal the actuality of a series of rapids on the St Lawrence that made this part of its route most difficult for all travellers to the west prior to the building of the first canals. The rapids were as beautiful as they were treacherous to navigate but the Long Sault near Cornwall dominated all. To see the entire flow of the river as turbulent white water for a distance of almost a mile was an experience never to be forgotten. The Long Sault and all the rapids above have now disappeared, drowned out by the impounding of water by the new dam and power house. The rapids below Lake St Francis, however, can still be seen. They are small indeed today since most of the flow goes down the power canal to Beauharnois but it requires only a little imagination to picture them as they were before man had interfered with the regime of the great river. We think of them today as beautiful but the early travellers probably had little time for thinking of their beauty as they struggled to get their small craft and their cargoes up their turbulent waters. Coming downstream, they had to concentrate all their attention on shooting the rapids that could be so navigated or in handling their craft with ropes down the swift water, when they did not have to unload and portage both cargoes and craft.

The first rapids on the journey upstream could be seen from Lake St Louis since, just as does the Ottawa a mile to the north, the St Lawrence enters the calm waters of the Lake through a series of rapids, generally known as the Cascades, but more specifically the Haystack and Split Rock rapids. Two miles upstream came the Cedar rapids around Cedar and some associated islands. Then followed a longer stretch of quiet water until the Coteau rapids were reached on the main river, again around a group of islands, a small part of the river here passing to the south of a much larger island, still called de Salaberry. We shall follow official practice in calling this the Soulanges section of the river when we come to see how it was bypassed by canals. Smooth sailing followed across Lake St Francis, and then through a singularly beautiful and relatively quiet section of the international part of the river up to where the city of Cornwall now stands, beyond which came the Long Sault. Just over 4 miles further upstream came the Farran Point rapids; then followed more than 9 miles of smooth sailing until the Rapide Plat was reached, one of the rapids that could be 'run' safely even by large vessels, except at low water, when in charge of experienced river men. After another 4 miles of pleasant sailing one came to the rapids at

Pointe aux Iroquois, Point Cardinal and the Galop in quick succession.

Once clear of the Galop rapid, sailing upstream was in the clear, through the beautiful Thousand Islands and so into Lake Ontario. We shall consider the canals built to circumvent these several rapids, collectively known as the St Lawrence Canals, in three groups. It will be seen that all the canals in the international section of the river were built on the Canadian shore even though always freely used by vessels of other nations, notably of the United States. It was Canadian initiative that resulted in the reallly remarkable investment represented by this St Lawrence canal system, remarkable when the fledgling character of the young country in its early years is recalled. To some extent the threat posed by the Erie Canal was responsible for the start of this extensive canal building. Fortunately, locating the canals on the Canadian side of the river took advantage of the better foundation conditions created by local geology, so that no unusual construction problems had to be faced. Nonetheless the early canals were fine examples of pioneer civil engineering.

The Soulanges Section

One of the first of the many contributions to the physical development of early Canada made by the Corps of Royal Engineers of the British Army was the building of the first small canals to overcome the rapids in this section of the river, to the overall direction of the Governor, Frederick Haldimand. As had to be noted in relation to the Lachine Canal, relatively little is known about these early works. They occupy a minor part in the overall story but the date at which they were constructed lends special interest to them. Starting in 1779, the Royal Engineers constructed small canals with wooden locks at La Faucille, Trou-du-Moulin, Split Rock and Coteau Rapids. The Coteau canal was the first to be tackled; it was 900 ft long and 7 ft wide with 3 locks, the depth of water it provided being no more than 2 ft. These dimensions will show that these early works were small indeed but the aid they gave to canoes and the *bateaux* then in general use on the St Lawrence must have been warmly welcomed, the depth provided being sufficient to float all normal vessels of the time. The Coteau Canal was finished in 1783, by which time work on a lock at Split Rock had been started; it was finished in 1783. Later the other two works were carried out, a canal 120 ft long without locks at Trou-du-Moulin, and a canal 410 ft long with one lock at Rapide à la Faucille just above Cascades Point. It is not surprising that, being inexperienced with ice such as forms on the St Lawrence, the Royal Engineers had to replace these last two works, after damage by ice, with a new Cascades canal, a major work 1,500 ft long with two locks, 120 ft by 20 ft at the lower end and guard gates at the upper entrance.[1]

This first Cascades canal was completed in 1805. The size of its locks and the increased depth of water made available (3½ ft) was indicative of changing patterns in sailing, Durham boats that would carry up to 35 tons being now in use. A second canal was later constructed at Coteau in 1817 with one lock, 104 ft long, 12 ft 6 in wide with a depth over sills of 4 ft, the canal being 400 ft long. These simple facilities served until replaced in 1845, tribute indeed to the good building of the Royal Engineers despite all the difficulties they faced. Quite naturally, there were increasingly strident demands for improvement of these aids to navigation. Indicative of the importance then being attached to water transport was a quite serious suggestion that the three sets of rapids should be circumvented by a new canal leading from the Lake of Two Mountains, across the level plain in between, to Lake St Francis. Surveys were actually carried out but when, after the union of the provinces of Lower and Upper Canada in 1841, a decision was finally reached as to the replacement of the 'Engineer Canals', a location on the south shore of the river (here Canadian) was selected. The bitterness of the arguments leading to this decision is difficult now to appreciate but one factor that weighed heavily in the discussions was the possibility of any canal on the south shore being more easily attacked from the United States than a new canal on the north shore.

H.H. Killaly was the chief engineer of the newly created Board of Works which had the responsibility for all public works in the newly formed Province of Canada, an eminent engineer whose vision is shown by the decision to make the locks on the new canal with 9 ft over the sills. It was known as the Beauharnois Canal, since it was located in what had been the seigniory of Beauharnois. Work started in 1842 and, despite the size of the project, was completed by 1845. The canal was 11.5 miles long, the entire length being a new excavation. It had 9 locks in order to provide for a total lift of 82 ft 6 in. The arguments about its location did not cease with its completion since the entrance from Lake St Francis, at its upper end, gave trouble from the start, with low water and boulders creating hazards to navigation. The trouble culminated in 1846 in the wreck of the *Magnet* at this location. Guide dams were then constructed in the river, utilising some adjacent islands, and some dredging was done in order to give safer depths of water. The slight rise in the level of Lake St Francis resulting from these remedial works necessitated the building of a long dyke along the south shore of the lake to prevent flooding of low-lying land. With these works completed, the Beauharnois Canal continued to give good service until the end of the century. The canal was used as a headrace for a small water power plant, discharging into the river at about the midpoint of the canal (where it came close to the river bank) and this proved to be one element in a drama that unfolded in the 1920s and 1930s at the time of the building of the great Beauharnois power station to be mentioned when we sail up the Seaway.

There was a 31 mile sail in smooth water from the western end of the Beauharnois Canal before the Long Sault was approached near Cornwall. The Cornwall Canal, built to circumvent the Long Sault, was first used late in the year 1842 and so was almost contemporaneous with the Beauharnois Canal. Before we deal with its construction and enlargement, however, we should finish the tale of the Soulanges section since, most strangely, when the decision was taken to procure a depth of 14 ft in all the St Lawrence Canals, the Beauharnois Canal was not enlarged. Instead, an entirely new canal was planned and built, but on the north side of the river, the older canal being abandoned except as a feeder to water power plants. The discussions and arguments that preceded this decision must again be left to the imagination but the decision was made and the Soulanges Canal constructed between 1892 and 1899 with 'fourteen foot locks', each 280 ft by 46 ft, but now only five instead of nine. At the east end there was a group of 3 locks, each with a lift of 23 ft 6 in; then, after a sail of 3½ miles, a fourth lock with a lift of 12 ft; and finally a guard lock at the upper end of the canal with a nominal lift of 1 ft.[2] The canal was, as can be seen, a continuous cut 14.67 miles long with the distinction of having its own small water power station about midway along, also discharging into the St Lawrence almost opposite the power station on the older canal. Except at its upper end the Soulanges Canal had no bends in its course, the long straight stretch giving a quiet sail along the flat plain that here lies between the Ottawa and St Lawrence Rivers. The old main road from Montreal to Toronto and the west ran parallel with the canal and close to it for over 10 miles so that the Soulanges Canal was probably the best known of the St Lawrence Canals to those travelling by road. This piece of road unfortunately gained an unenviable reputation for accidents in its more recent years as a main highway.

The Cornwall Canal

The Soulanges Canal served well until 1959 when the Seaway was opened. The 14 mile sail up the Soulanges Canal was followed by a pleasant 31 mile sail through Lake St Francis and up the first part of the international section of the river until the city of Cornwall was reached. Here the Cornwall Canal would have been entered just to the east of

the centre of the town and this would have been the original
canal, suitably enlarged to provide 14 ft depth. Eleven miles
long, and paralleling the river closely, it was provided with 6
locks, originally 200 ft by 45 ft with 9 ft over the sills. The
building of the canal started as early as 1834, in a way as a
response from the St Lawrence towns to the opening of the
Ottawa-Rideau route from Montreal up into the Great
Lakes. Progress was very slow at first, being stopped al-
together at the time of the 'troubles' in 1837 and not resumed
until after the union of the two provinces. Work started again
in 1842 and was so far advanced that in the late fall of that
same year, Captain Stearns managed to take the small
steamer *Highlander* through the canal and, without the aid
of the Williamsburg Canals, to sail up the remaining rapids
and on as far as Kingston. The arrival at Kingston of this first
vessel up the St Lawrence marked the turning point in the
fortunes of the Ottawa route. Thereafter, successive improvements of the St Lawrence route confirmed its priority,
climaxed in 1959 by the opening of the modern Seaway.

The original Cornwall Canal served well until the end of
the century. Upon the decision to standardise on a depth of
14 ft for the locks, reconstruction of the original locks on the
Cornwall Canal started. Inevitably, since navigation had to
be kept going without interruption during the open season of
navigation, this meant that the rebuilding would be a slow
job. It lasted from 1876 until 1904 but thereafter traffic flow
was steady and without interference. The Cornwall Canal
had the distinction of incorporating the large repair dock that
was so clearly necessary for a navigation system such as we
are considering. Between old locks Nos 16 and 17, and so to
the north of the canal proper, an area of about two acres was
available for berthing for repair work and other services.
Two canal-size vessels could be accommodated, draft being
limited to 12 ft, and also a considerable number of smaller
vessels. Access was through the old lock No 17. Dimensions
of the new locks were 270 ft by 45 ft and the standard 14 ft
draft, but with one peculiarity. Lock No 17 was only 43 ft 8 in

wide at its bottom and only 45 ft wide at its coping so that this lock was, if it may be so expressed, the 'bottleneck' of the St Lawrence 14 ft system. But the restriction that this unusual but slight diversion from standard dimensions caused was not too serious. Yet another unusual feature of the Cornwall Canal was the numbering of its locks, from 15 to 21 but not consecutively following on from those of the Soulanges locks![3] The sail through this canal was exciting, for those interested, since one got a splendid view of the Long Sault alongside and, on occasion, one could see the small white *Rapids Prince* with its load of adventurous passengers shooting this great and turbulent stretch of white water.

The Williamsburg Canals

The next rapid encountered on the St Lawrence was at Farran Point, about 5 miles upstream from the point at which the Cornwall Canal joined the river. Although there was rough water here, the experience of the *Highlander* showed that at low water small vessels could navigate past the rapids even going upstream, having no difficulty in coming through the rapids when downward bound. A canal, however, had to be provided for larger and loaded vessels upward bound. When built, it was the first of four rather similar small canals, eventually reduced to three. That at Farran Point was started in 1844, being opened for use in June 1847. It was only a little more than a mile long, with one lock to the earlier standard dimensions of 200 ft by 45 ft with 9 ft over the sills, as were the locks on the canals yet to be mentioned, when first built. The lift provided by the lock was only just over 4 ft, indicative of the minor character of the rapid which downward bound vessels always sailed through rather than losing time by locking down. After the decision to convert to 14 ft over canal lock sills, the Farran Point lock was rebuilt between 1899 and 1913, being doubled in size to speed up its operation, dimensions for its last 60 years of service being 800 ft by 50 ft with 16 ft over the sills. The extra 2 ft could not have been effective in view of the 14 ft used elsewhere on the

St Lawrence system. As one proceeded upstream, a sail in relatively smooth water for the next 9½ miles brought one to the foot of the Rapide Plat.

Although the total drop in the Rapide Plat was about 11½ ft, it was spread out over a fair length of river so that vessels downward bound could easily run the rapid. This they did regularly as long as the canal was in use, apart only from fully laden freighters at times of low water. These included the 250 ft lower lakers, their navigation through these several rapids being a matter of considerable skill but saving much time. The Rapide Plat Canal was almost 4 miles long, the full drop taken up in one lock at the town of Morrisburg. There was a guard lock at the upper end and the canal joined the river again at Flagg's Bay. The canal was built between 1844 and 1847, being opened in September of the latter year. It was rebuilt with standard lock dimensions between 1884 and 1904, without interference with regular shipping.

Another 4 mile stretch of easily navigable water followed, leading to the foot of the first of three rapids which here followed one another in fairly quick succession. First came that at Point aux Iroquois (a memorable name), the town of Iroquois being developed nearby; then a smaller rapid at Point Cardinal, now the location of the town of Cardinal; and finally the Galop Rapid. Since the total fall in all three rapids was only about 15½ ft, it is understandable that not only all downward bound but even upstream vessels could navigate them readily unless heavily loaded. Canals were needed, however, for upward bound freighters and so two small canals were originally planned and built, the first known as the Iroquois Canal, the second as the Galop Canal. They were constructed between 1844 and 1846-47, being opened for use in October 1847 and November 1846 respectively. One wonders if that difference in date of completion could have had anything to do with the mistake in the level of the upper sill of the lock on the Iroquois (the downstream) Canal. This error, well recorded at the time, is the only case of an error in engineering work that I have encountered in a study of all the canals of Canada. It did cause difficulties in 1848 and was the subject of considerable public discussion, so much so that the 'Junction Canal', a short section of waterway dyked to separate it from the main river, was built between 1849 and 1851, combining the two smaller canals into the Galop Canal as it was known for the next century. This gave the slight extra depth of water necessary to obviate the effects of the error.

The original locks were built to the usual dimensions but when the canal was rebuilt, for the usual reason, between 1888 and 1904 a rearrangement was made so that the entire drop was concentrated in the one lock at Iroquois with a lift of 15.46 ft at normal water levels. Advantage was also taken of the rebuilding to make this a double lock, just as at Farran Point, 800 ft long and 50 ft wide but here with just the normal 14 ft over sills. A guard lock near the head of the canal was also constructed to the usual 270 ft by 45 ft size. In addition a special river lock was built close to the guard lock through which downward bound vessels could pass back again into the river after bypassing the Galop Rapid only, making the rest of their way down the river channel. If the canal was used for its full length, the canalised length was 7.36 miles, but with another 2½ miles at the upstream end protected by dykes to facilitate navigation in what was a somewhat difficult little section of the river. Beyond the west end of the North Channel dyke, all was clear sailing for the next 229 miles to the entrance to the Welland Canal.[4]

Use of the St Lawrence Canals

After their final rebuilding, the St Lawrence Canals had standard minimum dimensions for their locks—270 ft by 45 ft with 14 ft normal draft over sills. These dimensions were critical. They determined the limiting dimensions of vessels from European countries (and others) that sailed across the Atlantic and then up the St Lawrence and into the Great Lakes. This was a common practice long before the Seaway was planned, going back at least to 1860. More important, however, was the fact that these dimensions determined also the size of lake vessel that could sail from Lake Ontario to the port of Montreal.

Before the present Welland Canal was finished in 1931 this meant the determination of the size of lake vessels in general but we shall see that the new Welland Canal permitted the largest lake vessels then sailing the upper Great Lakes to come down into Lake Ontario as far downstream as Prescott (see p 174). This small port is just upstream of the head of the Galop Canal and assumed considerable importance between 1932 and 1959. It remains a pleasant and important town even though its great days as a transhipment port have gone. A large grain elevator was constructed here by the Government of Canada for storage of grain brought down from the head of the lakes by large upper lakers (as the great vessels used to be known, up to 600 ft long). The grain would then be discharged from the elevators into lower lakers, these being vessels specially built to conform to the controlling dimensions of the locks on the St Lawrence (and Lachine) canals. There was a fleet of almost two hundred of these stubby, efficient but hardly beautiful vessels before the Seaway made them obsolete. Some sailed down the Gulf of St Lawrence and others into the upper Great Lakes, on their lawful occasions, but most of them provided a shuttle service between grain elevators on Lake Ontario and the port of Montreal where grain was stored in a battery of vast elevators before shipment by ocean vessels across the seas. Usually 253 ft long, the lower lakers each carried 2,800 tons of iron ore or 106,000 bushels of wheat in three holds. They deserve this brief tribute to their faithful service.[5]

As we considered the various improvements and enlargements that were effected to the successive small canals in this section of the river, there was no real opportunity for recounting all the discussions and inquiries that prefaced the carrying out of all these works. Committees and Commissions had this critical waterway under review for almost a century. Many were the arguments about how it should be

developed, all quite apart from the great debate as to ultimate development of the St Lawrence route or that up the
Ottawa. As we know well, it was the St Lawrence route that
achieved Seaway status but, in the overall picture, it is at
first puzzling to find that the '14 foot canals' (as the final
enlargements were known) remained in service for so long
while other major ship canals of the world were being developed for much larger vessels than those that could proceed up the St Lawrence to the Great Lakes. The delay is the
more puzzling when it is realised that, in their final years, the
14 ft canals were operating at full capacity. The reason for
the delay was the long drawn-out international discussion
about the development of the St Lawrence, both as a major
seaway and for power generation, in the international section. This complex story will be summarised when we come
to consider the Seaway of today but it must be thus briefly
mentioned here to explain why the bottleneck of the 14 ft
canals, good as they were, persisted for three quarters of a
century, until 1959.

Despite the bottleneck, the little canals with the lower
lake fleet of canalers managed to provide a most efficient
transport system, especially for the wheat coming from the
prairies on its way to hungry nations around the world. The
economics of this canal system must first be briefly mentioned since it might be thought that the necessary double
transhipment of wheat coming down from the lakehead
would have made the St Lawrence route uneconomic. Early
competition was from the Erie Canal, later rebuilt as part of
the New York State Barge Canal system, but in recent years
the St Lawrence system could meet this competition quite
successfully. Shipment of wheat from the prairies by way of
the Churchill route through Hudson Bay naturally gave low
ocean freight rates, despite high insurance costs, but the
open season at Churchill was so short that the Bay route
could be regarded as a supplement rather than as a com-
petitor to the St Lawrence route. When Vancouver developed as a major ocean port, wheat became a major part of

its freight business. For shipments to the Far East it was naturally unrivalled but all shipments bound for Europe from Vancouver had to sail down the West Coast in order to pass the Panama Canal, thus increasing shipment costs. In the days of enthusiasm for railways in Canada, there were great hopes that wheat would be shipped in large quantities by rail. Grain elevators were erected even at small ports on Georgian Bay, Lake Huron, with the hope that wheat coming thus far by water would then be transhipped to rail and so utilise railways all the way to Montreal and Halifax.

These high hopes the success of the St Lawrence Canal route completely sabotaged, apart from winter shipments of wheat to Halifax and Saint John. And although wheat declined from being one quarter of all Canada's exports in 1910 to about one eighth in the 1960s, the total volume exported increased so that it remained a good indicator of transport economics. We cannot take final leave of the St Lawrence Canals without giving thought to the good service they rendered to Canada, the United States and indeed to the trade of the world, through their more than a century of service, despite the limitations that their small locks imposed upon the traffic they carried. The large fleet of lower lakers became familiar to most Canadians living anywhere near the St Lawrence. Far from beautiful, they were serviceable and seaworthy. Their departure, as 'uneconomic' with the opening of the modern Seaway, left a gap in the overall picture of Canadian shipping but it will be long before they are completely forgotten.

CHAPTER 11

The Welland Canal

It is not often that a canal has been misnamed, but although the Welland Canal might well have been called the Niagara Canal, it was designated differently when first authorised in 1824, and the Welland Canal it has remained. This is the canal that crosses the Niagara peninsula between Lake Ontario and Lake Erie, with a length of about 28 miles. It parallels the Niagara River, this being the name of the connecting river (instead of the St Lawrence). Its lift of 327 ft is accounted for by the drop in the world-famous Niagara Falls and the rapids both above and below this splendid cataract. The present canal is the fourth, successive enlargements having generally taken the form of entirely new canals although parts of earlier canals were used to some degree. The steady improvement of the Welland Canal over a period of 150 years is an important chapter in the history of Canadian civil engineering, even as it is also a reflection of the increasing requirements seen by successive Governments of Canada to be essential for the country's waterways. There are no big 'ifs' in the story of the Welland Canal, as we have seen to be the case with some other Canadian canals, and only one big question mark. This is why and how the building of the fourth (present) canal was authorised as early as 1912 and whether this was in any way associated with the proposed Georgian Bay Ship Canal. But we must place this question in its proper context and so first consider the early start of this justly famous waterway.

From the time that 'The Falls' were first seen by a white

man (Father Hennepin in 1678), and the St Lawrence route had been shown by the Indians to Joliet (see page 119), the need for some passage around the Falls and the associated rapids had been obvious. To begin with, the simple Indian portage trails were probably first used, but a wagon road from Queenston, at the foot of the downstream rapids, to Chippewa above the upstream rapids, was one of the earliest of such roads in Canada. So important was this portage seen to be that it was guarded by a French military post as early as 1687. It passed into British hands in 1759. Shortly thereafter at capstan incline was constructed on the west side of the Niagara River to the top of the escarpment, up which it was possible for small *bateaux* to be hauled. From the top a wagon road had to be used for the six miles to Chippewa, the toll being the substantial sum, for those days, of £10 New York currency. Despite this aid, the human labour necessary to get boats and freight up that big hill and over the intervening land between the two landing points can only now be imagined although contemporary accounts show clearly that the portage was well used.

The idea of a canal must have been early in men's minds. One of the earliest records of this is contained in a report of an emissary of the French King Louis XIV, Clerambaut d'Aigremont, who commented to his royal master that 'if M. de la Mothe knows a means of doing so [connecting the two lakes], I think he is the only man in the country who does. But, my Lord, even if it were true that a communication between Lake Ontario and Lake Erie could be made, it could only be done with very great expense.' How right he was! But the water connection was made, in the first instance by a private company, one of the few Canadian canals that was first built by private interests and not by government.

It was in the state of New York, however, that the first proposal for a canal connection between the Great Lakes and the sea was advanced. A natural route from Lake Ontario to the east, leading to the Hudson River at Albany, existed in the Mohawk valley. The first serious proposal for a canal using this route appears to have been made in 1807 by Jesse Hawley. From the time of this first suggestion, all US advocacy was for a canal linking the Hudson River with Lake Erie, and not with Lake Ontario; thus the name given to this great pioneer venture, Erie Canal, was not a misnomer. Despite the increase in length that extension to Lake Erie involved, traffic from the upper Great Lakes using such a canal would be clearly destined for New York with no possibility of its diversion to Montreal, as could have been the case if it were routed through Lake Ontario. This was the prime argument of De Witt Clinton, the great advocate and eventually the builder of the Erie Canal. Despite all his efforts, it was not until after the close of the War of 1812 that New York State finally decided to proceed. The building of the Erie Canal commenced on 4 July 1817.

Developments leading up to this auspicious event were naturally known in the Niagara region. Proposals were advanced for a 'Niagara Canal' paralleling the Niagara River on the US side of the border but this idea was strongly opposed by the merchants of Buffalo who did not wish to see any of their trade diverted down the St Lawrence. In Upper Canada, however, no such objections could be raised and so it was on the Canadian side of the border that the first really serious proposal for a 'Niagara Canal' was advanced and where, eventually, the canal was built. William Hamilton Merritt is the man credited with first proposing the construction of the canal, the idea of such a venture having come to him, according to his son, when 'riding along the Niagara River to the Chippewa ferry' in the year 1812.

Merritt was born in the Carolinas in 1793 but was brought up to Canada when he was three months old. At the age of 15 he made a journey to St John, New Brunswick, by way of the St Lawrence and this journey probably gave him his life-long interest in travel by water. It was not, however, until 1818 that he made a rough survey of a possible route for a canal to feed water to his mill on Twelve Mile Creek from the Welland River. The creek was one of a number of short streams, all with similar names, which drain off the escarpment into Lake Ontario. The Welland River, on the other hand, lies above the escarpment so that this early survey had to include for passing water through a local high ridge near the crest of the great slope and then down the slope to the level area making up the Lake Ontario shore. On the survey, Merritt was accompanied by George Keefer and John De Cew, neighbouring mill owners, their names still well known in Canada. The actual start of work on the Erie Canal changed the initial idea of a water feeder from the Welland River to the Creek into the more daring idea of a canal through which small boats could be sailed from one lake to the other.

The influence of the Erie Canal on the ideas of Merritt and his friends is clearly reflected in the wording of a petition addressed to the Legislature of Upper Canada—the first of several—on 14 October 1818. Here is the relevant extract: 'The grand object of the American people appears to be opening a navigation with Lake Erie, which design our canal, if effected soon, would counteract; and take down the whole of the produce from the Western country.' This was a prophetic statement since, although the rebuilt Erie Canal still serves well New York State, the great bulk of modern traffic from 'the Western country' does come down the Welland Canal on its way to the sea by way of the St Lawrence and Montreal. There was to be much discussion, however, before this far-sighted proposal of Merritt and his friends won official approval, including (surprisingly) considerations of defence.

Much of the land-based fighting of the War of 1812 took place in the Niagara region so that it is not surprising to find that any proposal for a canal close to the Canadian-US border would be viewed askance by British military authorities. An alternative proposal was therefore seriously advanced for a canal between Burlington Bay, in the north-west corner of Lake Ontario, and the Grand River which flows into Lake Erie about 35 miles west of the mouth of the Niagara River. Although the proposal did not win eventual support, it was one of the several factors which caused a delay in granting permission for the start of construction on the Welland Canal until 1824. The small naval base established at the mouth of the Grand River later led to a change in plans for the feeder canal that linked this river with the new canal.

Following many negotiations and much discussion, Merritt and his friends presented another petition to the Governor, Sir Peregrine Maitland, on 18 May 1823, requesting permission for the building of a canal 'from the mouth of Twelve Mile Creek to the River Welland.' On 19 January 1824 the Legislature of Upper Canada passed an Act incorporating the Welland Canal Company, the three men already mentioned being amongst the incorporators, the authorised

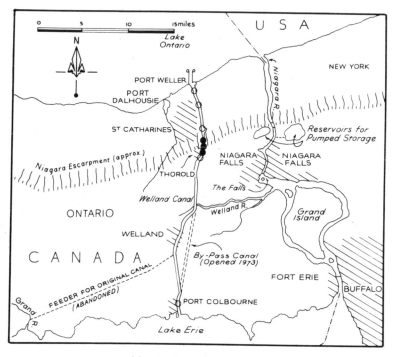

Map J: The Welland Canal

capital being only £40,000, to be raised by shares of £ 12-10-0 each. In March of that year, Merritt was in Quebec City raising the first actual cash, the initial £1,000 being quickly subscribed; he came back with about £10,000 promised.

The accompanying sketch plan shows the general lie of the land so that the strange plans for the start of this great work can be followed with its aid. If the years of construction be kept in mind, 1824-1829, then the timid planning of the originators will be the better appreciated. They were courageous in thinking that a connection between the lakes could be made at all but the engineering advice they received was not all that it might have been. The original idea was to use the waters of the small Welland River (or Chippewa Creek as it was also known) as far as possible, with a tunnel through the high ridge of land that lay athwart the route to the edge of the escarpment, then to have a channel to the top of the big slope down which *bateaux* would be lowered, and correspondingly raised, by means of a 'Railway'—as it was described—leading to a landing place on Twelve Mile Creek which would then be followed for navigation into Lake Ontario.

This scheme was not followed as planned but the idea of going only as far as the Welland River was a leading feature of the canal when first opened for use. It was early decided to use a series of locks, instead of a railway, for passage up and down the escarpment. Bad ground conditions gave trouble in excavating the tunnel. A 'Deep Cut' was substituted but this meant a higher water level in this part of the canal. Two locks had therefore to be built in the Welland River, to lock *up* vessels on their way down to Lake Ontario. A feeder canal to supply water for this upper reach of the canal was essential. This was achieved by excavating a long cut to the mouth of the Grand River. Since the normal water level in Lake Erie, and hence at the Grand River, was 12 ft above the level of the Welland River (this being the drop in the Niagara River from its mouth to the entry of the Welland River), a suitable supply of water was thus made available in addition to providing a second canal outlet to Lake Erie.

The original locks were of timber, no less than thirty-four being necessary for getting up and down the escarpment. They were 100 ft long and 22 ft wide and were supposed to have a depth of 7 ft 6 in over sills although in fact 7 ft was about the maximum clearance obtained. Even from this capsule description of the works involved, it can readily be appreciated that the original £40,000 of capital soon proved to be inadequate. The Canal Company was beset by financial worries almost from the start of its work, regular appeals having to be made to the Legislature for more assistance. The Legislature refused to take more stock in the company and so, as a last resort, the Imperial Government was appealed to. They provided a loan of £55,555. The Duke of Wellington may have had some influence in this, in view of his interest in and concern for canals in relation to the defence of Canada. Papers in the archives of Ontario show that the Duke personally invested in 25 shares of the company; Mr Merritt must have been a good salesman on his visit to England.

Despite all difficulties, including mistakes by contractors, the canal was finally completed from lake to lake. On 20 November 1829, two schooners—the *Ann and Jane* and the *R.H. Broughton*—were taken through the canal from Lake Ontario to Lake Erie by way of the Welland River. A small boy who saw these two small vessels passing through the new canal lived to the great age of 94 and was able to tell fellow members of the American Society of Civil Engineers (of which he was a past president) about this famous day when the society met in Ottawa in 1913, just as work on the fourth Welland Canal was starting. This was Thomas C Keefer, one of Canada's great pioneer engineers, whose father had dug the first sod for the canal on 30 November 1824. A memorial cairn at Allanburg marks the spot at which this historic event took place. The courage of the builders of this pioneer canal can still be admired when it is realised that at the time it was built, there was only one small steamer on Lake Erie, the upper Great Lakes being still relatively unused. The entire fleet for all commerce above Niagara comprised no more than forty vessels and only two were larger than 100 tons. Although the Rideau System omitted towpaths in favour of steam haulage, horses were to pull freight through the Welland Canal. This decision was derided by writers of the time. One of these has described how, on the first Welland Canal, 'any horse was good enough for towing; the old, the poor, the halt, and the blind were therefore procured for this purpose, and as they were killed by the work in a few weeks, it was soon found to be true economy to pay £30 and £40 each for the best which could be procured, and the class of animals now employed for this purpose are not to be surpassed anywhere.'[3] Traffic increased, some US vessels even bypassing the Erie Canal at Buffalo, using the Welland Canal and Lake Ontario as far as Oswego, where they entered the Erie system.

Despite all the favourable early estimates, the small canal had cost over £300,000 by 1830 and so there were many demands for an inquiry into the conduct of the company. A commissioner was appointed by the Legislature in 1830 but his report did not help very much. The directors of the company knew that a direct connection to Lake Erie (instead of using the Welland to the Niagara River) was essential, and this was built; so there were more demands for help in raising funds. Authority was given for raising more stock since this further improvement to the canal was clearly essential. Eventually, the total cost was £407,855 of which £288,000 was provided by governments as loans or by stock purchase. It is small wonder that the entire project was described by a writer in 1865 in these words: 'a wilder, more ill considered scheme than the one originally put forth, one showing more ignorance and recklessness on the part of the projectors, it is scarcely possible to conceive.'[4] But the canal was built and did operate, even if inefficiently. The main complaints were made about the conduct of the canal company, culminating

in thirty specific charges against the directors by William
Lyon Mackenzie, one of the leading public figures in this
turbulent period in Canadian history. Eventually it was decided that the canal must be taken over by the government.
An Act to authorise this was passed by the Legislature of
Upper Canada on 5 July 1840 but it did not become law until
5 July 1841 at the first session of the Legislature of the united
Province of Canada. Indicative of the feelings of the day is
the fact that the report to the Upper Canada House of
Assembly of the committee that looked into Mackenzie's
charges is a volume of no less than 575 pages! It exonerated
all those associated with the works but revealed many serious inefficiencies.[5]

The foregoing brief summary of this tangled history has
been given as some indication of the difficulties that surrounded the early days of this important Canadian canal,
similar but perhaps rather more unusual than the early development of some of the other canals that have already been
described. After 1841, as can be imagined from some of the
histories already related, the newly created Board of Works
assumed control and proceeded to do a good job in re-
habilitating the original works. New locks were first constructed at both ends of the canal to facilitate the entry of the
new steamships then coming into use; they were built of
masonry to enlarged dimensions of 150 ft by 26 ft 6 in with 9 ft
over sills. Eventually the other locks were replaced by
masonry structures, the total number of locks being reduced
from 40 to 27. An additional terminus was built on Lake Erie
and this was connected with the feeder canal which had a
new branch built to Port Maitland at the mouth of the Grand
River. It is, today, almost impossible to understand this
emphasis on the use of the feeder canal and other connections to the Grand River, through what is now pleasant rural
country. The remains of these little used feeder canals must
stand as yet another memorial to 'canal fever'. The work of
rehabilitation was naturally spread over a number of years,
the 9 ft depth between the junction with the feeder canal and

Port Colborne, the exit on Lake Erie, becoming available only in 1850. So rapid was the increase in the use of the canal that the normal navigating depth was increased to 10 ft, starting in 1853, by the device of raising the embankments forming the canal and the tops of the masonry lock walls.[6]

The reconstructed canal served an increasing volume of traffic for the next thirty years when it, too, had to be improved. Most of the locks of this second canal, however, are still in existence. It is interesting today to compare their puny size with the great locks of the present (fourth) canal. The southern section of this second canal is still in use, but as a channel for waste water and for the discharge from the De Cew Falls power station. This has been possible since, as we shall shortly see, the northern terminus of the canal was changed from Port Dalhousie to an entirely new port for the fourth canal. The feeder canal from the Grand River continued to operate through this period even though low flow caused increasing problems. This followed from the gradual development of agricultural land in the Grand River basin and the associated removal of forest cover. The situation became so serious in the 1930s that one of the first river conservation projects to be carried out in Ontario was the construction in 1939-40 of the 75 ft high Shand Dam on the Grand River, solely for the purpose of increasing low river flow. Use of the Grand River to feed the Welland Canal, however, had then long since ceased although it was still so used until 1883 and to a small extent for some years thereafter.

In keeping with the improvements effected in the St Lawrence canals in the last two decades of the nineteenth century, the Welland Canal was also greatly enlarged in its capacity at the same time. A new route was chosen for that section of the third canal between Port Dalhousie and the foot of the escarpment, the town of Thorold being the point at which the climb up the escarpment started for this 1883 canal. There were some favourable topographic features that made this route up the escarpment an improvement over that used for the first two canals. Once at the top of the big hill, however, the third canal followed generally the line of the second, directly to Port Colborne on Lake Erie. As originally completed in 1883 the third Welland Canal had a depth of 12 ft over the lock sills but further improvements were made and more dredging so that by 1887 a 14 ft draft was available throughout, this being the agreed upon depth for all the canals in the St Lawrence system.[7]

Comparison of the dates just given with those already cited for the periods during which the St Lawrence canals were reconstructed to give a through 14 ft depth, with standard locks at least 270 ft by 45 ft, will show that the development of the complete 14 ft system was a prolonged affair, covering generally the last twenty years of the nineteenth century. This lengthy construction period is no reflection upon Canadian skill in building, even at that time, but rather an indication of the financial problems of the times, the indecision about the future of canal traffic, and competition for public funds from railway construction. There was probably lacking any broad conception of the position that canals would come to occupy in the handling of bulk traffic such as wheat from the prairies on its way to the ocean. Larger vessels were already in use on the upper Great Lakes for the movement of iron ore from the mines of Minnesota to US steel mills in cities on the shores of Lake Erie in particular. Similarly large vessels could bring wheat down from the head of the Great Lakes but only as far as Port Colborne, at the Lake Erie end of the Welland Canal. It was not, however, until 1908 that a grain elevator was ready for use at Port Colborne to assist with the transhipment of wheat from these large upper lakers to the lower lakers which could then sail directly from Port Colborne through the Welland, St Lawrence and Lachine canals to Montreal. Here corresponding transhipment arrangements had to be built. And at Port Colborne, an adequately protected harbour had to be developed in order to provide protection for vessels waiting to use the government elevator.

All these works were completed, however, in the first few years of the new century and the 14 ft canal system was well launched on its thirty years of service—only thirty for the complete system, because the fourth Welland Canal would be officially opened in 1932. Thereafter, the 14 ft system started with the St Lawrence canals, the large upper lakers being then able to sail down to Prescott on the international section of the St Lawrence. It was in the first decade of the century, as was noted on p122, that the rivalry between the St Lawrence route and the Ottawa route to the Great Lakes reached its peak of acrimony—and difficulty for political leaders who had to make the necessary decisions. One decision that government leaders do not appear to have made, however, was to start surveys (about 1906-08) for the construction of an entirely new Welland Canal, built to accommodate the largest vessels then contemplated for use on the Great Lakes. So far as one can gather from available records, this decision was made by the engineer in charge of the Welland Canal on his own initiative, or at least without specific Parliamentary authorisation. It is not now profitable to spend time on such a detail of canal development. The significant thing is that the surveys were made; plans were drawn up; estimtes of cost were prepared. When, therefore, the Conservative Government of Sir Robert Borden soon after it had assumed office (p124) faced the problem of deciding between the two routes, there was available to it not only the complete scheme for development of the Ottawa route, in the D.P.W. Report of 1908, but also a correspondingly complete presentation for the construction of a major ship canal to replace the third Welland Canal.

It is generally thought that the decision in favour of the Welland Canal, and so of the St Lawrence route, was based at least in part on the estimated time for the completion of the two projects. It is difficult today to understand how such a comparison could have been regarded as valid, when construction of the new Welland Canal would still leave the bottleneck of the 14 ft St Lawrence canals, while the Georgian Bay Ship Canal would have provided a deep waterway all the way from Montreal to Lake Huron. Again, however, such speculation is of historic interest only, since the Welland Canal won the favourable decision on the basis, so it is averred, that its construction would take only four to five years as compared with a decade or more for the Georgian Bay project. A small start on the fourth canal was therefore made in 1913. The outbreak of World War I slowed up operations almost immediately, work being suspended completely in 1917 and 1918. It re-started in 1919, again slowly, but it was not until 1931 that the new canal was ready for use, and not until the summer of 1932 that it was opened officially. This great ceremony was held during the course of the British Empire Economic Conference that was convened in Canada. The Governor General, the Earl of Bessborough, declared the Canal open for use, in the presence of a most distinguished gathering, on 6 August 1932.[8]

When the Welland Canal is visited today, either on land or by water, all these problems, frustrations and difficulties are forgotten as one admires—even today and despite so many other more recent engineering achievements—the vision, skill and courage of the small group of engineers who first conceived this great ship canal. They designed it so well that it fits almost perfectly as an integral part of the modern St Lawrence Seaway with no substantial change to its structures or layout, and they built it well despite the interruptions in the work. The conception of a canal with locks well over 800 ft long and 80 ft wide, with lifts of almost 50 ft, is even today remarkable; in the early nineteen hundreds it was visionary indeed. Those responsible may have been inspired by the renewed interest then being shown in the Panama Canal as it was taken over by the United States. I always regret that, as a young man, I did not have the sense to ask about this when I met the man who was chief engineer of the Welland Canal throughout its entire construction, Alexander J. Grant, then a grand old man in his seventies.

A native of Banffshire, Scotland, Mr Grant joined the

then new Department of Railways and Canals in 1886, being engaged at first on the building of the Soulanges Canal. He supervised the improvements at Port Colborne between 1903 and 1906, probably hearing early discussions of the great new canal then being dreamed about. After being in charge of the Trent Canal from 1906 to 1919, he returned to the Welland Canal as chief engineer, serving until this entire project was finished and in operation before retiring in 1934. He was supported by an able and devoted group of assistant engineers, some of whom went on to distinguished careers in other places. They placed Canada, and indeed the United States, in their debt. It had been decided to make an entirely new canal up to and over the escarpment but thereafter to follow generally the line of the third canal to Lake Erie. A new port had therefore to be built on Lake Ontario; it was called Port Weller. A new canal was excavated in a fairly direct line to the foot of the escarpment along Ten Mile Creek, three locks bringing the water level up by almost 140 ft. The escarpment was mounted by what may properly be called the famous flight of three twin locks, giving a further rise of almost another 140 ft, with Lock No 7 just over the crest. This brought the canal essentially to the level of Lake Erie, the remainder of its route being through a channel that was originally far from straight, as far as the guard lock at Port Colborne, whose small lift varies with the level of Lake Erie. Normally, the total lift from Lake Ontario into Lake Erie is 327 ft, achieved in just 7 locks with the addition of the slight lift in the guard lock. With 6 of the locks 859 ft long and 80 ft wide, the fourth Welland Canal is in striking contrast to its first predecessor with its forty little timber locks.

Since we shall be sailing up this fourth Welland Canal when we make our imaginary voyage up the Seaway, it will be convenient to describe its main features at that time, in order to avoid repetition. Suffice to say here that it has been in unbroken service since 1932, carrying a steadily increasing volume of traffic. It has been excellently maintained with a succession of minor improvements down through the years. Only one major change has been made, this being the construction of a new 'cut-off' around the town of Welland, a new section of canal with a straight alignment, 8.3 miles long, opened for use in 1973. Control of the movement of all vessels in the canal has also been modernised, as we shall also see when we come to sail up this original part of the Seaway.

This, then, is the Welland Canal of today, assuredly one of the world's great canals, its continued and increasing service a tribute to the sound planning of fifty years ago and to the excellence of its original construction. As a reminder, however, of the way in which the canal of today developed from its predecessors, this chapter may well come to a close with a summary tabular presentation of the four Welland Canals which have been so briefly described:

	1829	*1845*	*1887*	*1932*
Length in miles	27½	27½	26¾	27½
Locks: number	40	27	26	8
Locks: length in feet	110	150	280	859
Locks: width in feet	22	26½	46	80
Locks: depth in feet	8	10	14	30

CHAPTER 12

Locks at Sault Ste Marie

Once vessels leave the Welland Canal they have a clear sail of over 200 miles to the west end of Lake Erie, or less if they are bound for one of the ports on this lake such as Cleveland. The route to the head of the lakes takes a sharp turn to the north when this west end is reached, and the channel is then up the connecting rivers between Lakes Erie and Huron. The Detroit River extends from Lake Erie to Lake St Clair, a roughly circular body of relatively shallow water about 25 miles across. The city of Detroit lies at the head of the river which bears its name and extends for some miles along the west shore of Lake St Clair. From this lake to Lake Huron the connecting river (still really the St Lawrence) is called the St Clair River; it has a course almost due north and south and is about 38 miles long, the municipalities at its mouth being Sarnia in Canada and Port Huron in the United States, the international boundary running approximately up the middle of the two rivers and directly across Lake St Clair between them. Then follows another 200 mile sail to the north-west end of Lake Huron where the St Mary's River enters this lake from Lake Superior, the St Lawrence having here yet another of its four alternative names. No canal works have been necessary in this 500 mile length of waterway but only considerable dredging to develop and maintain the necessary navigation channels.

The St Mary's River is about 70 miles long between Pointe Iroquois at the exit from Lake Superior to the end of St Joseph Island where it enters Lake Huron. The drop in level between the two lakes varies between 19 and 22 ft, depending on the levels of the two lakes. Most of the fall is concentrated at one location where the falls on the river gave their name to the twin cities of Sault Ste Marie, in Ontario, Canada, and Michigan, USA, respectively, about 14 miles to the east of Pointe Iroquois. This location, colloquially 'The Soo', has therefore been one of great importance from the days of the first French explorers and fur traders, since portaging of canoes and their loads was always necessary. A French mission was established at an early date (in 1667) and there has been continuous settlement here since that time, never large but always important. At the height of the great days of the fur trade, the North West Company constructed a canal on the Canadian side of the falls. This was in 1797-1798: a replica of this eighteenth century canal may still be seen today. It was a small affair, with one wooden lock measuring 38 ft by 9ft wide with just 18 in of water over the sills. It was built, however, to serve the canoe traffic and so was adequate for this purpose, and greatly welcomed by canoe travellers. A British visitor in the first years of the nineteenth century (George Heriot) described the facility in a book that he wrote on his *Travels through Canadas* in these words: 'At the factory {of the N.W. Company} there is a good canal, with a lock at its lower entrance and a causeway for dragging up the bateaux and canoes. The vessels of Lake Superior approach close to the head of the canal, where there is a wharf: those of Lake Huron to the lower end of the cascades....'[1]

The little canal was partially destroyed on 20 July 1814 in one of the forays, even in this isolated location, that distinguished the War of 1812 but it was quickly rebuilt more substantially than before and continued to give good, even if limited, service until the end of the canoe era. With the coming of steamships and so an increase in trade on the Great Lakes, a larger canal was an obvious necessity. Indicative of the advance of settlement in the United States is the fact the the first moves for a larger canal at The Soo were made on the Michigan side of the St Mary's River. As early as 1837 Governor Mason of the new state of Michigan authorised surveys for a canal; these were completed. Since the Congress of the United States would not give any assistance, the state of Michigan awarded a contract for the building of the new canal in September 1838 but since part of the area to be used was a military reservation, the contractors were thrown off the site by the US Army in May 1839. Congress continued its opposition to the idea of a canal. It is interesting today to read that no less a man than Henry Clay, in opposing the project, described it as 'a work beyond the remotest settlement in the United States if not in the moon'. The 'plank and tramroad transfers' had to continue. It was not until 1852 that the President signed a bill authorising a grant of 750,000 acres of land. A company was organised: locks were designed, two in tandem, each 350 ft long and 70 ft wide with 13 ft over sills. Work started in June 1853; the first vessel went through the locks on 19 April 1855. From the first, the locks were freely used by vessels of both the United States and Canada.[2]

Improvements were made in 1871 and 1881, after which year a depth of 16 ft was available over the sills. The US Corps of Engineers of the US Army began their long association with the Sault Ste Marie locks at this time, a connection that still continues; two of today's locks carry the names of the first two generals, Poe and Weitzel, in charge of works at The Soo. Both of the early locks have now been superseded but before the first modern US lock was built, the Government of Canada constructed the Canadian lock on their side of the international boundary. The Canadian lock and short approach canals are still in use, paralleling the four newer US locks. Started in 1887, the Canadian lock was completed by 1895. St Mary's Island, a long narrow island parallel with the flow of the river, facilitated construction even though it involved an excavation 7,294 ft long, generally 150 ft wide at the top and about 140 ft at the bottom, with a depth in the upper approach canal varying from 21½ to 24 ft. The lock is 900 ft long and 60 ft wide with 18.25 ft over the sills.[3] One year later, the United States completed the (old) Poe lock which had 21 ft of water over sills, its approach channel having been deepened to 25 ft.

By the turn into the new century, a start at the design and construction of special vessels for bulk trade on the Great Lakes had been made. With the growth of the steel industry, the carrying of iron ore began to loom large, in addition to the steadily increasing trade in wheat. The 'locks at the Soo' (to use the colloquial expression) had a corresponding growth. This was all on the United States side of the border. Although the US locks naturally belong to the story of United States canals, one really cannot consider the canalisation at Sault Ste Marie without considering all five locks that exist today. They have always been operated together; together they handle a far larger volume of traffic each year than any other canal system of the world, and this despite the limitation to a nine to ten month open season of navigation.

In summary, therefore, there are now on the US side of the border at St Mary's Falls, the David and Sabin locks each with 23 ft of water over sills, 1,350 ft long and 80 ft wide; and the MacArthur and (new) Poe locks, each with 31 ft of water over sills, the former 800 ft long and 80 ft wide, the latter 1,200 ft long and 110 ft wide. Upstream of all five locks is also an international railway bridge with nine fixed spans and a bascule bridge over the north (Canadian) canal and a vertical lift bridge over the south canal with a clearance under it of 124 ft. Upstream of the group of locks is a fixed highway bridge with the same clearance beneath its spans which cross the respective canals. The concentration of interest in this relatively small area can well be imagined even from these summary statistics.

It is only at the Canadian lock that one can see the restoration of the tiny original lock of the North West Company. Somewhat naturally all the larger vessels use the newer locks on the US side. But one can stand not far from the reconstructed little old canal and watch the largest lake vessels go through the new Poe lock. The contrast is striking and one would have to be insensitive indeed not to appreciate the significance of this vivid reminder of the progress that has been made in shipping on the Great Lakes in the short space of less than two centuries. As traffic continues to increase, it seems more than probable that the Canadian lock will be reconstructed with larger dimensions before too many more years have passed. This would be in keeping with the cordial international co-operation that has distinguished, now for so many years, the development and operation of this great route into the heart of the continent, co-operation now sealed by the joint construction of the St Lawrence Seaway and Power Project. To this great project we must now turn our attention as the climax of our study of the canals of Canada.

CHAPTER 13

The St Lawrence Seaway

'Our canals were not built for Canada, but for the *valley of the St Lawrence*; we ought therefore to "club together" with our neighbours, on the opposite side in order to place this noble outlet in the most efficient state, by giving it as large a support as possible.'[1] These words appear in a prize essay, 'The Canals of Canada,' written by Thomas C. Keefer (already mentioned appreciatively in this book) and published in 1850—and the italics are his. In the year 1895 a Deep Waterway Commission was established jointly by Canada and the United States to investigate the idea of a deep waterway into the Great Lakes. In 1932 a treaty between the two countries covering this long discussed development was signed by the Prime Minister and President respectively but the US Senate failed to ratify it. Even under the urging of every President of the United States of America from President Wilson on, and particularly of President Roosevelt, the Congress continued to refuse to take the necessary action. Canadian views during these years are perhaps best expressed in some words of Stephen Leacock who was, it should be remembered, Professor of Economics at McGill University as well as a writer of memorable books: 'To hell with economics; it's a magnificent conception, and has got to be built.'[2] But it was not until 22 November 1957 that the first vessel passed through the newly completed Iroquois lock, marking the beginning of the end of the construction of this project.

Why should it have taken so long to get this magnificent

183

concept to the construction stage, one of the truly great canals of the world, now looking so 'natural' in its pleasant setting as the usual scars from construction operations gradually disappear? Numerous volumes much larger than this one have been written on varied aspects of this perplexing question. And this is a book on canals and not one on politics. A summary answer only can be given, therefore, but such a brief conspectus is essential if the Seaway is to be fully enjoyed and appreciated. Once the development of power in the International Rapids section had been associated with the construction of the deep waterway, the whole concept took on new and increased significance but so also did the opposition. The factor that swung one or two critical votes in the US Congress was the need of US steel mills on the Lakes for large quantities of the iron ore that was ready to be mined in Quebec-Labrador, the transport of which up the Gulf of St Lawrence and then into the Lakes making a deep draft seaway vital if transport costs were to be economical. When this fact was coupled with an earlier suggestion, accepted by both countries, that tolls should be charged on all vessels and freight passing through the Seaway (even though this meant a reversal of Canadian policy unchanged since 1903), the battle was won.

Against this general background, let us now take a quick overall look at the highlights of the strange and prolonged story of this political battle—for that is what it turned out to be. We must first remind ourselves that there are three main routes from the sea to the Great Lakes. The first is up the Gulf of and River St Lawrence and then up the Ottawa Waterway into Lake Huron. As we have seen, this was the main route to the west for two centuries. It was canalised only as far as the present city of Ottawa, the die being cast in 1911 against its complete canalisation as an all Canadian seaway to the Great Lakes. The power available in its many rapids has now been developed in a series of great water power plants, in only one of the impounding structures for which has a ship lock been installed. The second route is also up from the sea to Montreal and then up the St Lawrence River into Lake Ontario, the route we shall examine in some detail. The third approach to the Great Lakes was by way of New York, up the Hudson River to Albany, then turning to the west up the Mohawk River valley which led through an unusual geological gap in the mountains to a portage over to Wood Creek, the Oneida River and so down to the level plain to the south of Lake Ontario, the old 'Mohawk Trail'. This was the route followed by the Erie Canal (as it was first called), now a part of the New York State Barge Canal system. Originally built in 1817-1825, it is still in active use, naturally after considerable rebuilding and improvement.

With the completion of the 14 ft improved St Lawrence and Welland Canals by the end of the century (see p 159), the advantage swung to the Canadian route and this quite naturally led to demands for improvement of the rival route down the Mohawk valley. After much public discussion, an entirely new Barge Canal was authorised by the State of New York, started in 1905 and completed in 1918, generally on a new alignment although always close to the original canal. But it was still a barge canal, with locks only 328 ft long, 45 ft wide and with a depth of water over sills of 12 ft in general but 13 ft from the Hudson River to Oswego, to which port a branch canal had been constructed giving connection with the east end of Lake Ontario. The canal is of course crossed by many bridges—generally of the swing or bascule type, since the topography around the canal is not conducive to giving with fixed bridges the necessary overhead clearance for normal types of lake vessel. The new Barge Canal proved to be efficient in operation to the extent of its limited capacity, and is still an important inland waterway but it did not provide real competition to the Canadian 14 ft canals which carried an increasing amount of traffic. Small as were these Canadian canals, they carried not only an extensive traffic of lake vessels but an increasing number of ocean-going vessels of the requisite limited size. By the year 1954, near the end of their use, these canals were served by 14 overseas shipping

lines, involving 120 vessels, there being no less than 225
overseas sailings out of the port of Chicago in that year
alone.

Such was the increase in traffic on the lakes, however, that
almost by the time that the 14 ft system was complete it was
recognised as inadequate. Locks at Sault Ste Marie and the
new Welland Canal eventually (by the thirties) took care of
the movement of large upper lakers within the area of the
Great Lakes themselves but the link to the St Lawrence at
Montreal remained the '14 ft bottleneck'. Large volumes of
iron ore were being brought from mines of the Messabi
Range at the head of Lake Superior both to US steel mills on
the Great Lakes and to two Canadian steel plants, at The
Soo and at Hamilton on Lake Ontario. Coal in vast quantities from the mines of Pennsylvania was being moved on
the Great Lakes from shipping points on Lake Erie. And
wheat, the most bulky material of all, was moving from the
mid-west of the United States and the Canadian prairies to
Atlantic ports both by rail and by water even though this
involved inevitable transhipment, either on the Erie Canal or
at one of the Canadian ports on Lake Ontario (see p 160) into
lower lakers for transport to Montreal and further transhipment there into ocean vessels.

This, in broad outline, was the situation when, in 1895,
Canada and the United States jointly appointed a Deep
Waterways Commission. It was directed to study all possible routes between the Atlantic and the Great Lakes, so that
competition between the Mohawk route and the St Lawrence was still in the air. The Commission reported in 1897 in
favour of improving access up the St Lawrence. From this
same date organised opposition to any such improvment also
started; the lobby had begun. Those opposed included the
railways, quite naturally, Atlantic coast ports, ports in
Texas, and surprisingly mid-west cities; all combined eventually into the National St Lawrence Project Conference—a
grandiloquent title that was explained by its sub-title *A
Nation-Wide Organization in Opposition*. And opposition

they surely did mount, with complete success for half a century. The arguments in opposition, all based on natural self-interest, can be imagined but even imagination boggles when faced with the suggestion that if the Seaway were built it would permit warships of the Royal Navy to penetrate the Great Lakes to the detriment of United States defence measures. The opposition was focussed in what has been described as the most powerful lobby that ever existed in Washington, for the purpose of making sure that this great public project should not be built. As late as 1950 the Chairman of the Conference (from industry) declared publicly with regard to the proposed legislation that his motto was 'They shall not pass'. But eventually they did.

Initial studies made by the Deep Waterways Commission were of navigation requirements only but the development of water power at Niagara Falls in the last decade of the nineteenth century pointed the way to another factor that was to assume increasing importance with the years. As early as 1901 a small privately owned water power plant was built at Massena, New York, using a small amount of water from the St Lawrence and part of the drop in the International Rapids section, a plant subsequently enlarged and in use until 1959. In 1918 and again in 1921 official applications were made on behalf of important industrial groups for permission to develop privately all the power in the International Rapids section. These were not granted. As we shall see, public power organisations would be given the authority to develop all this power, but the general public interest in the St Lawrence as a whole had already been indicated in 1909. It was in that year Canada and the United States signed a treaty, which was approved by the Congress, establishing the International Joint Commission. In 1914, after President Wilson's administration had reaffirmed the treaty, the Commission was asked specifically to study the navigation and power possibilities of the St Lawrence. Thereafter all studies of the international section of the great river involved both power and navigation.

It will surprise British readers to know that the development of this power was not to be carried out by the senior governments of the two countries involved. In both countries, however, there are three levels of government—federal, provincial or state, and municipal—and through legal interpretations of long-standing statutes control over natural resources has been placed, in each country, in the hands of the appropriate second level of government. The Hydro Electric Power Commission of Ontario was well established in its distinguished career as one of the great public utilities of the world by the year 1932 when the Government of Canada finally, and after a strange series of internal events, allowed Ontario to start planning for the use of St Lawrence power. Just one year before this, the State Legislature of New York had established the New York State Power Authority to be responsible for its share of the great block of international power. This action added still further to the opposition forces since private power companies and major suppliers of coal saw their interests being interfered with. But the decision to proceed with public power development, once made, was not changed.

A joint board of engineers had been engaged on detailed studies of possible schemes of development, under the IJC, and agreement was reached in 1932 on a two-stage power project with the necessary locks and other navigation works to provide for a 27 ft deep waterway into the Lakes. Based on this agreement, a treaty for the construction of the entire project was signed in 1932 but failed to win the necessary two-thirds majority for ratification in the US Senate (although it did receive a simple majority vote in favour). Despite all the efforts of President Roosevelt during the thirties, no progress was made; the opposition stood its ground. The years of World War II gave no opportunity for consideration of such a major peacetime project and so it was not until the war approached its close that the Seaway again became a topic of wide public concern. It was in 1946 that the suggestion of charging tolls, and making the Seaway a self-liquidating proposition, was first advanced. The suggestion met with favour and probably made the first dint in the staunch opposition to any suggestion of the Seaway being built, but still not enough to get things moving.[3]

Far away in the wilds of Quebec-Labrador, however, iron ore mines were being opened up, prior to full development, that were to be the ultimate deciding factor. It was in 1949 that the construction of the 330 mile Quebec North Shore & Labrador Railway was started from the (then) tiny port of Sept Iles on the north shore of the Gulf of St Lawrence up on to the Labrador plateau where the new mines were situated. This signalled the reality of vast quantities of good iron ore soon to be available at water level on the Gulf and within easy sailing distance of the steel plants on the Great Lakes—if the Seaway were available to make such transportation economical. Correspondingly, the power situation had changed dramatically during the years of war, especially in Ontario. Ontario Hydro was desperately short of power in view of the great increase in the province's industrial capacity as a result of war production and corresponding increases in all types of demands for power. All of its economically available major water power sites had been developed. Diversions of Arctic water had been made into Lake Superior to increase flow down the St Lawrence and so the power potential of the river. By international agreement, new power plants were built at Niagara Falls on both sides of the Niagara River. All this, however, was not enough. The frustration of those responsible for power policy can well be imagined if thought be given to the two million horsepower still undeveloped in the International Rapids section of the St Lawrence, so close to the heartland of Ontario.

It was not at all surprising, therefore, when Canada made it quite clear to the world, and to the United States in particular, that if the United States would not join with it in the development of the St Lawrence, Canada would build the new deep water navigation facilities on its own and at its own expense. The Canadian prime minister, Mr St Laurent, at a conference in the White House in Washington, told President Truman in September 1951 of the plans that his country had already made, and made it quite plain that Canada was serious in its intention. I can recall so clearly being told at the time by engineering friends in the United States that Canada must really be joking to have made such a suggestion. Never was a nation so united or so adamant in its intentions, as was shown by the speedy passage by the Parliament of Canada, later in 1951, of the St Lawrence Seaway Authority Act which established the St Lawrence Seaway Authority as the Canadian agency responsible for constructing 'all such works in connection with such a deep waterway as the Governor in Council may deem neces-sary....' Assent to the new legislation was given in January 1952. This action was followed by an application by the Canadian Government, under the terms of the 1909 treaty, to the IJC for permission to develop the power in the interna-tional section of the St Lawrence, jointly with the United States. An Order of Approval was issued by the IJC in October 1952 authorising the construction of works neces-sary for the generation of St Lawrence power. The end of the long fight was in view.

Once the IJC had issued its order, the Power Authority of the State of New York applied to the Federal Power Com-mission for the necessary licence which was granted in 1953 for a 50-year period. Opponents immediately contested this in the courts but the US Court of Appeals denied a rehear-ing. This decision was then appealed to the Supreme Court of the United States for review but the Supreme Court denied this. At last the way was clear for the power de-velopment to proceed, these legal details being given to confirm what has already been indicated about the nature of the opposition to the Seaway and power development. Meantime, progress was being made in Congress at long last. Some leading and influential senators changed their views and votes, with the result that on 12 January 1954 the matter finally reached the floor of the Senate, after a long protracted committee stage. The corresponding debate in the House of Representatives followed emergence of its bill from its committee stage on 22 February of the same year. The final bill authorising construction of the Seaway from the United States side became law on 13 May 1954. It was known as the Wiley-Dondero Act in honour of Senator Wiley and Representative Dondero of Wisconsin, who had guided the legislation to final victory over all opposition. The way was now clear for construction to commence but naturally there had to be discussions between the two countries as to 'who would do what' and how co-ordination would be achieved. This was confirmed by an exchange of Notes, following close consultation. Compromises had to be made, perhaps the main one being that the navigation locks in the international section of the river would, for the first time, have to be built on the United States side of the river.

Agreement was reached quickly on all points, however, with the result that on 10 August 1954 at the site of the Canadian power house near Cornwall the sod-turning ceremony was performed by the Right Honourable Louis St Laurent, Prime Minister of Canada, the Honorable Thomas E. Dewey, Governor of the State of New York, and the Honorable Leslie Frost, the Premier of Ontario, in the presence of a distinguished international assembly. Less than five years later the official opening of the entire project was performed close to the same spot by Her Majesty Queen Elizabeth II of Canada and the Vice-President of the United States in the absence of the ill President Eisenhower. And the Seaway had been in full use since the opening of the navigation season on 25 April 1959; both power stations had been delivering power since 1958. This four to five year period for construction may be compared with the ten years spent in building the Suez Canal with no locks, and the 24 years required for completion of the Panama Canal, as also with many years taken for the building of the original Canadian St Lawrence canals and even for their improvements. In some ways, the operation may be justly regarded as the greatest peacetime construction project ever carried out, certainly at such speed, and in entire harmony between the two participating countries.[4]

From what has already been said, it can be realised that construction was not only an immense project overall but the responsibility for it was spread between four major agencies—the two power authorities, the St Lawrence Seaway Authority (Canadian) and the St Lawrence Seaway Development Corporation (the corresponding United States agency) established in 1954 under the Wiley-Dondero Act. The two senior governments naturally had a vital interest in the progress of the work as did also the governments of Ontario and New York. And everything was under the ever-watchful eye of the International Joint Commission. It would be tedious to list even the more important of the other agencies involved directly in the complex construction operations. Merely as an example let there be mentioned Canadian National Railways and the province of Quebec through its great public utility, Hydro Quebec.

The main double-track line of CNR between Montreal and Toronto had to be relocated for a distance of about 20 miles where it ran over land that would be flooded once water was impounded by the power dam. In itself this was a major operation in view of the importance of this line and the heavy traffic it carries but this was only one of the many such ancillary operations, all of which had to be co-ordinated with the progress of the major works. Hydro Quebec was deeply involved since it now owned the great Beauharnois water power station that had been built by a private corporation between 1929 and 1932. From the map on p 206 it can be seen that this project, now generating almost two million horse-power, had involved the construction of a very large power canal roughly parallel to the old navigation canal (which it naturally dwarfed). The political repercussions of this development form a peculiar chapter in the history of Canadian power development but of interest in this context is the fact that the Government of Canada at the time very wisely anticipated the building of the Seaway and the use of the power canal as a navigation canal. In the design of the bridges crossing the canal, arrangements were therefore made for the ultimate replacement of central fixed spans by movable spans, and this saved a great deal of work on this part of the Seaway 25 years later.

The size of locks to be adopted was naturally a crucial question, one that had long been under discussion in view of the economic implications of the limitations that the size of locks would place on ocean vessels wishing to use the Seaway. The decision was made to use the same dimensions as those already used for the locks on the Welland Canal, namely: 800 ft long (765 ft usable) and 80 ft wide with 30 ft of water over sills. Channels, however, would be dredged initially only to a depth of 27 ft as a minimum. This logical solution to a difficult problem was not without its critics but, as we shall see later, the size of locks does not seem to have interfered too greatly with the use of the Seaway by ocean-going vessels. The other great design decision was to develop the power available in the International Rapids section in one stage only, with an international power house at Barnhart Island, close to Cornwall, Ontario, and Massena, New York. One of the factors that is believed to have influenced this decision was the advance in the engineering uses of soil as a construction material which permitted the construction of the necessary retaining dykes (really earth dams) upstream of the power station, on both sides of the river, up to heights of 75 ft with confidence in their long-time performance. By skilful design measures, it was possible to utilise as much as 83 of the 92 ft drop between the normal level of Lake Ontario and Lake St Francis, 125 miles downstream. Each part of the vast power house contains 16 generators each with a capacity of 75,000 hp, so that the combined station can produce 2,400,000 hp when all machines are running. The entire output was needed almost as soon as the job was finished!

With these great decisions made, design work proceeded

in innumerable offices, there being as many as 500 professional engineers engaged at the peak by the four main agencies alone. In summary, the works involved the construction
by Canada of two locks to bypass the Lachine rapids at
Montreal and a new canal on the south shore of the river up
into Lake St Louis and two locks to get from this lake into the
Beauharnois power canal. Then came the power house,
control dam and two US navigation locks in the International Rapids section; another special control dam, and a
lock to bypass it, at Iroquois, 50 miles above the power
house, for the control of Lake Ontario; and all the associated
dredging, bridge reconstruction, railway, road and town reconstruction. Ten miles of new canal had to be built and
eighteen miles of specially constructed dykes upstream of
the power house. If I went on to give total quantities of
material excavated and used for construction, the figures
would be so great as to be beyond normal comprehension.
Let it just be said that at one time no less than 22,000 men
were at work, between Montreal and Lake Ontario, all
working to a wonderfully co-ordinated programme, and
under a master contract (in Canada) which ensured no
strikes, after one unfortunate experience at the start of the
work. The total cost for the two power developments was
about $600,000,000, shared between the two authorities.
Corresponding total cost of the Seaway was about
$470,000,000, about 70 per cent of which was spent in Cana-
da, but all costs were, by international agreement, shared on
a fifty-fifty basis including the original cost of the fourth
Welland Canal which now formed so vital a link in the
Seaway.[5]

  Before we begin an imaginary journey up the Seaway from
Montreal, readers may like to know about just a few of the
more interesting details of construction. At Montreal, for
example, four major bridges connecting the island of
Montreal with the mainland to the south had to be reconstructed to give necessary clearances. First was the Jacques
Cartier bridge, carrying a vital and major highway, opened in

1930. The Seaway was to go under the south approach section made up of deck spans, an increase in clearance of 80 ft being essential. All the spans were therefore jacked up progressively, the piers for the key span being raised 50 ft in small increments and with no interference with the heavy traffic in its six lanes; few of those who used the bridge knew that it was being raised a little each day! The key Seaway span was replaced, in a carefully planned operation, by a through truss span which gave the additional 30 ft necessary, the operation taking less than six hours (the only period during which the bridge was closed) while the spans were slid off and on to adjacent falsework respectively. Next came the Victoria bridge carrying not only highway traffic but the two main lines of Canadian National Railways into the city from the east and south. The ingenious solution reached here will be best described when we travel up the Seaway. Another large highway bridge had to be reconstructed and raised, and two new lift spans inserted into another important railway bridge carrying the main lines of the Canadian Pacific Railway across the St Lawrence and into Montreal. At the Beauharnois locks, it proved desirable to carry an important main road going west beneath rather than over the locks and so a tunnel was excavated in solid rock before work on the lock structure above it was started.

The most critical engineering operations were naturally centred around the control of the Long Sault (rapid) at Cornwall. The main river channel was blocked off by a major cofferdam above the head of the rapid, the flow of the river being thus diverted into the channel south of Barnhart Island. A graceful curved floodway dam equipped with twenty-four 50-ft wide gates was later built across this channel as the permanent control for all water not being used in the power house. Construction of the main cofferdam resulted in the drying up of the bed of the Long Sault, thus revealed as a mass of boulders and broken rock overlying the bedrock, an awesome sight expecially for all who had ever sailed down the great rapid. It is said that the last captain of the S S *Rapids Prince*, when taken to see over what he had guided his sturdy vessel down through the years, averred that he would never have sailed it had he known what was beneath him. The construction of another large cofferdam downstream of the site of the power house permitted the construction of this vast structure to proceed 'in the dry'. The combined power house is 3,300 ft long, the main generators being covered with removable hatches, the building proper being below this unusual roof structure and yet still 167 ft high. Built in two separate sections by the two power authorities, the combined power house has the outward appearance of an integrated structure. This is symbolic of its building. On a lovely summer day at the height of construction I was kindly shown around the work by my friend, the director of the project, for the benefit of the chief scientist of India, then a guest in Ottawa. As we stood at the top of the construction work our visitor asked where was the dividing line between the two countries. In explaining that it was not then marked in any special way, the director said that the two organisations were working so closely together on the two parts of the one structure that there had not been any difficulties or serious arguments from the start. Our Indian guest listened in amazement and said in words full of meaning: 'How can such things be?'

All this was a part of the Seaway construction as well since the dam formed by the power house structure impounds the river above it into what is now called Lake St Lawrence, at approximately the level of Lake Ontario. Two navigation locks were necessary to raise vessels from the channel coming up from Lake St Francis into the new lake. By international agreement these were located on the US side of the border. Their construction involved a very large excavation job since a new channel had to be formed through what had been largely dry land. Approaches to this new route were excavated (or improved) by dredging, but excavation for the two locks and the two-mile channel between them was carried out 'in the dry'. Another of my privileges at the time was to be able to visit the excavation for the (upstream) Eisenhower lock when this had been carried to bedrock through over 100 ft of three different types of glacial till, naturally most carefully studied by expert geologists, the leader of whom was our guide. It was a singularly moving experience to stand on smooth glaciated bedrock that had not been exposed to daylight for possibly a million years and to know that within a matter of weeks it would again be covered up, but this time by man-made concrete, never to be seen again in the foreseeable future. Access to all these operations was another vital ancillary work, several special bridges being built and a tunnel beneath the Cornwall Canal to give access to the Canadian power house site.

When the water was eventually raised behind the power dam, it flooded 38,000 acres, 20,000 in Canada. Flooded land in the United States was cultivated and important but, unlike that in Canada, it did not include any important settlements. Seven Canadian villages, all long established, and one third of the town of Morrisburg, were so located that their sites would be flooded. Ontario Hydro had the responsibility for the major rehabilitation job that relocation involved, one of the most complex ever carried out. Three new towns were planned, close to where the new shore-line would be. Well laid out, the new towns included many new houses and service buildings but also 525 houses which were moved, each in one operation, from their old foundations to new concrete foundations at the new sites. This was done with the aid of two 'house-movers', large automotive frames that could fit over a complete house, one capable of lifting 100 tons, the other 200 tons. The associated social problems can well be imagined especially as eighteen cemeteries had also to be relocated, but all was done well, and on schedule. Buildings that were not moved were destroyed. Ontario Hydro kindly permitted some buildings to be destroyed by burning, after being completedly instrumented, in order to give information for fire research never previously available. Driving around the deserted village of Aultsville in order to select the buildings to be used for research was an eerie experience especially when one of my (then) colleagues in a sepulchral voice would say: 'I think we'll burn that one'. Eleven buildings were thus burned, most successfully, with results of much benefit to the advance of fire protection engineering.

Not only were new towns laid out in Ontario but pleasant parks along the new waterfront created by the flooding. These have been steadily improved in the years since completion of the main project, marinas having been developed for the steadily increasing use of Lake St Lawrence by pleasure craft. So also in the United States, Barnhart Island and adjoining land has been converted into the Robert Moses State Park, a place of real beauty. The US part of the power station has also been named after Robert Moses, the director of its power project, the Canadian section being correspondingly named the Robert H Saunders station, after the Chairman of Ontario Hydro who contributed so much to the St Lawrence project. At the two ends of the power house, unusuallly good facilities have been built by the two authorities for the convenience of visitors. A visit to either one is warmly commended to all who have occasion to be anywhere near Cornwall-Massena, well over 100,000 visitors now coming to the Canadian power house alone every year. Films, working models and splendid viewing areas give the visitor a real feel for the importance of the power side of this great joint development. Visitors are correspondingly welcomed at all the locks, parking and other facilities being well arranged and even a two-tier 'grandstand' provided alongside the Eisenhower lock so that visitors can watch all phases of locking through a vessel. These pleasant attractions naturally followed the successful completion of the construction project, all of which was completed on schedule, some major works being finished well ahead of scheduled dates which were originally regarded by some as impracticable.

The lock at Iroquois was the first major work to be completed, this being necessary in order to facilitate the initial changeover from the 14 ft canals, which had naturally to be kept in full and regular operation, prior to the flooding of the power pool. The adjoining control dam was built by the New York State Power Authority, the lock by the Canadian Seaway Authority. On 22 November 1957 the Canadian Minister of Transport operated the control panel to admit the first vessel to this lock, and so to the new Seaway, this being the Canadian lighthouse tender CGS *Grenville*. The lock was ready for full use from the start of the 1958 navigation season. By 4 July of the same year the two US locks were ready for use. Three days prior to that, however, the climax of the whole job was reached when, at 8:00 a.m. on 1 July 1958 (Dominion Day, Canada's national holiday) a button pressed jointly by the two chief engineers of the authorities detonated a charge of 30 tons of nitrone which breached the cofferdam above what had been the Long Sault. A crowd of over 50,000 watched this critical operation, the breach allowing the pent-up waters of the river above to flood the previously dried-up area, right down to the power house. All went as planned, the excellent progress and scheduling being shown by the generation of the first power only four days after water first reached the power house. It was now possible to close off the Cornwall Canal and this and other co-ordinated operations proceeded as planned with the result that regular through transit of the entire St Lawrence Seaway started with the opening of the 1959 navigation season on 25 April 1959, power production having been steadily increased as more machines were completed. The official opening took place in June of that year.

CHAPTER 14

A Journey up the Seaway

As a fitting conclusion to this all-too-brief survey of the canals of Canada, let us take our imaginary journey up the St Lawrence Seaway, now that we have reviewed its long and complex history. We could make the journey in a small pleasure craft. One of my friends has made the journey in a 30-foot sloop (fortunately equipped with a small auxiliary engine). He found the arrangements for pleasure craft at each lock satisfactory but was somewhat awed by the comparison between the size of his craft and the locks he used, especially as he watched one of the large lakers come through a lock. The Seaway, however, was designed to accommodate large ocean-going vessels as well as pleasure craft, even to the most diminutive, and lakers of all sizes. Let us, then, make our journey through the medium of these pages in a 500 ft ocean freighter (of 15,000 tons), as I have been privileged to do in fact. Our floating home for ten days will be boarded at one of the riverside wharves in the east end of Montreal Harbour, to which it had sailed across the Atlantic from Europe and where it has unloaded some of its general cargo as well as some of the many containers stacked on its deck. Most of this now-usual deck load is intended for ports on the Great Lakes, the necessary handling equipment having added appreciably to the jibs, spars and cables now necessary on the vessel itself, despite the fine container unloading equipment at all larger Great Lake ports. Some of our journey through the Seaway proper will necessarily be at night, as will be realised when times of transit are given.

For convenience, however, the journey will be described as if it were all done in daylight; such is its fascination that sleep is inevitably sacrificed by passengers.

When all is ready for sailing, lines will be released with no ceremony and the vessel will move slowly upstream into the St Mary's current which can so clearly be seen sweeping down through the main part of the harbour. Slowly and cautiously a course will be set across the current in order to approach the south shore of the river where the unassuming entrance to the Seaway will soon be seen, between the south shore and St Helen's Island. As the canal channel is entered, interest will be divided between the magnificent skyline of the city, dominated by Mount Royal, the buildings that still remain on the island from Expo '67, that magnificent internatonal exhibition, and the approach to the Jacques Cartier bridge, towering over the channel as it gives the necessary 120 ft clearance above water level. The heavy traffic carried by this bridge will clearly be seen as also will be that along the roads running parallel to the canal on the south shore and close to it, some having been built on new land reclaimed with excavation from the canal. This continuing road traffic on our port bow will continue throughout most of the 20 mile length of the canal until we enter Lake St Louis. A bright red light ahead shows clearly the entrance to the first (St Lambert) lock. With a vessel obviously locking down, our speed will slow almost to a stop since the downward bound vessel must be well clear of the lock before the two vessels may pass, as they can quite readily, the canal width being a minimum of 200 ft although wider at many locations. Without stopping at the berth below the lock, we shall enter it very slowly after the entrance signal has changed to green.

Lock gates are the usual mitre type of welded steel, 50 ft high and 48 ft wide, electrically operated. Even before we watch the locking operation, our attention will be attracted by the unusual bridge arrangement here, there being steel vertical lift bridges at both ends of the lock. Long explanation can be eliminated by reference to a photograph of this lock on page 137 in which its location near the end of the Victoria bridge will be seen. This bridge, still using the original piers of 1860, carries a double track of Canadian National Railways with at least 120 trains a day, and a busy double lane highway. Neither rail nor road traffic can be interfered with; the double bridge arrangement was the ingenious, costly but most satisfactory solution to the problem. With the downstream bridge up as we enter the lock, all traffic will be seen using the upstream bridge, diverted from its straight-through (original) route by switching, and traffic lights with barriers, respectively. Once the rear gates have closed behind us, the rear bridge is quickly lowered, traffic re-routed, and then the upstream bridge raised to permit us to leave the lock. So rapid is the filling of the lock, less than five minutes being necessary to lift us the normal 15 ft, that the bridge movements will be complete just in time for the safety wire-rope fenders to be raised (there is one at each end of every lock), the gates opened and our clearance given. When our deck is level with the lock wall, we see an interchange of men in mufti, one carrying a small bag. These are our first two pilots, one responsible for bringing the ship from the harbour into the St Lambert lock, the other now responsible for our course as far as the first US lock. Pilotage services throughout are for vessels of foreign registry, masters of lake vessels serving as their own pilots.

Once clear of the lock, we now have a slow but steady sail up the canal for 8 miles to the Côte Ste Catherine lock, the canal formed by an embankment on the north side and by the new shoreline of reclaimed land on the south. The second lock is similar to the first, less than ten minutes being necessary to lift our vessel the normal 30 ft, but this time with no complications from adjacent bridges. The next pair of bridges are four miles above the lock (see p 195) and after we have passed under them we have 2½ miles more in canal before entering Lake St Louis. This 6½ mile stretch of canal is located along the original shoreline of the river and the upper part parallels the magnificent Lachine Rapids of which we get a superb view. The day will come when the power so obviously available in this fall of the entire flow of the St Lawrence will be developed by Hydro Quebec. All Seaway planning was done with this eventual power project clearly in mind. The sail across the calm waters of Lake St Louis is a pleasant interlude, with good views of the Montreal suburban developments on both shores and, at the start, of the Caughnawaga (Indian) village, to avoid which the canal was routed out into the river again for the final short section.

The approach to the Beauharnois locks gives us one of the most spectacular vistas of the whole voyage. Turning to the south of Ile Perrot at the west end of Lake St Louis, we will steam slowly past les Iles de la Paix to port and Fortier Pointe to starboard just before passing the immense power station building through which most of the flow of the St Lawrence now comes from the power canal above, generating about 2 million hp as it does so, capacity of the station being 2,161,000 hp. Ahead of us we can see the relatively small Vaudreuil Rapids (p 51), one of the channels through which the Ottawa River joins the St Lawrence. On the port bow we see more rapids, these being the Haystack Rapids at the foot of the Soulanges section of the St Lawrence, now displaying a mere shadow of their former grandeur but a reminder still of the dangers and hazards of the original St Lawrence route to the lakes. Slowly turning to port, we shall approach the first of the two locks built near the west end of the power house, the two separated by 3,600 ft. Study was made of building the two locks in tandem, and also of constructing one major lock with a total lift of 82 ft, but the solution reached was found to be most favourable for a number of important reasons. Assuming no delays, it will take only 6½ minutes to lift our vessel the 41 ft in each lock. A typical schedule for a complete locking operation appears on p. 00

We have been happily assuming no delays as if ours is the only vessel sailing up the Seaway, an assumption with no validity at all! Such is the traffic on the Seaway that until all the St Lawrence locks are cleared, one is rarely if ever out of sight of another vessel either ahead or astern. Close to any one of the locks it is not uncommon to see two vessels downward bound and two or three on their way up river. Each of the locks has often many hours of unbroken operation so that the three shifts of lock staff are kept busy, the Seaway naturally operating around the clock. With traffic congestion so frequent at every lock, the mooring berths upstream and downstream of each lock are well used. Equipped with bollards and capable of berthing two large vessels, these wharves have a distinction all their own.

Typical locking schedule for the Beauharnois locks

Ship entering lock	15 to 25 minutes
Close fenders, then gates	3 minutes
Fill or empty lock	7 to 8 minutes
Open gates, then fenders	3 minutes
Ship leaving lock	8 to 12 minutes
	Total 36 to 51 minutes

Although the lock staffs handle all lines at the locks, mooring lines at the berths have to be handled by ships' crews, all vessels having to be equipped with the necessary swinging booms for lowering and retrieving the seamen necessary to take care of the lines. Once on the wharves, these men may have to wait there for an hour or more. Some industrious chap once took a can of paint and a brush down with him and loyally painted the name of his ship on the surface of the wharf before anyone could stop him. His example was quickly followed, with the result that the world's greatest display of graffiti—all very sober, mainly ship's names and origins—is to be seen on the mooring wharves along the Seaway. It is naturally only from an upper deck of a large ship that one gets the full view of this decorative feature. My friends who have not travelled the Seaway write this off as 'just another of Legget's tall stories' but the photograph on p138 will perhaps convince readers that this is no figment of the imagination but an unexpected and interesting feature of the Seaway. Of unusual interest are the different scripts to be seen so well painted on the asphalt surface—English, Arabic, Japanese, Greek, Cyrillic—vivid reminders that this is a *Seaway* carrying ships from all over the world into the heart of the North American continent.

Some of these we shall see amongst the constant procession of ships we are passing. Some are small and nondescript; others have been clearly designed especially for the Seaway service, approaching the 730 ft maximum length permitted, their decks festooned with necessary loading and unloading equipment. Publicity has made its imprint here, many—probably most—of the foreign vessels having the name of their line painted in large letters on the sides of their hulls, to be seen not only by those on other vessels but by the hundreds of travellers walking or driving along the roads that parallel the Seaway in so many places. One familiar name, *Manchester Liners*, will be seen on the hulls of relatively small but trim new vessels, quite unlike the larger Manchester liners familiar on the North Atlantic. The small vessels have been specially built to convey freight from Great Lakes ports to ocean-going Manchester vessels at Montreal, despite the extra transhipment. I know nothing of the considerations that led to this development but it confirms to a degree what many critics said prior to the building of the Seaway, namely that the slow transit speed and high insurance rates for ocean boats on the Great Lakes would militate against any great use of the Seaway by large ocean-going freighters. Many such freighters are regularly using the Seaway but the tonnage carried by ocean vessels is still not much more than one third of the total. This is still almost equal to the total tonnage carried by the Seaway in its first year of operation (1959), a good indication that the seemingly constant stream of vessels that we are passing does indeed indicate a great volume of traffic, a total of almost 60 million tons in 1973.

We are still only at the Upper Beauharnois lock, however,

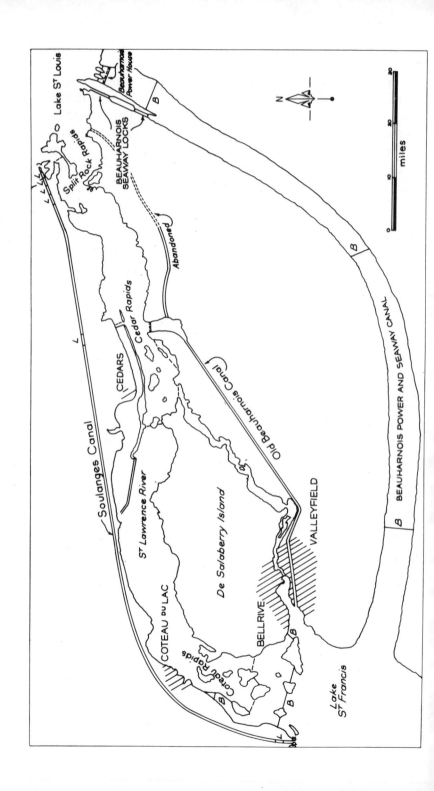

and so we should get further on our way before continuing these inevitable digressions. We could well have been held here by the Seaway authorities since, if fog is bad on Lake St Francis (for example) as it is on some mornings in spring and fall, all Seaway traffic is stopped, by radio control from St Lambert, until visibility again reaches the necessary level for safe navigation. One has only to see three or four large vessels waiting at a lock to realise how essential it is that *all* vessels are stopped when fog interferes, frustrating though this is to the masters of vessels in sections that are in the clear. Once we pull away we quickly enter the power canal, now dredged to give the necessary 27 ft depth, its embankments so well covered with vegetation that it is difficult to realise that we are in an artificial canal formed by embankments for much of its total length of 15 miles. We pass through the open life-spans of three bridges (p 193) and sail into Lake St Francis after passing the rear of the busy industrial city of Valleyfield. As we enter the lake, by looking astern we can see the course taken by the original St Lawrence, now bridged but still attractive to the eye. We now have a pleasant sail of 44 miles, first across the lake, the channel well buoyed, and then up a similarly well marked channel with many attractive islands, some of them completely lined along their shores with the summer cottages so widely used by North American city dwellers. We entered international waters when leaving the lake but there is no sign to tell one this, in keeping with the truly international character of the whole project.

We shall see, somewhere along this part of the route, almost certainly, maintenance work in progress—repairs of buoys or lighthouses, or minor dredging operations. If we are lucky, we may even pass the S L S *Hercules*, the Canadian Authority's self-propelled floating revolving crane. It has a capacity of 275 tons and so can lift any one of the lock gates on the Seaway, should this be necessary, rotate with it and lay the gate on the rear of its pontoon; if both gates of a pair have to be removed, it can repeat the process. At about mile 40 from the entrance to Lake St Francis, we shall see ahead of us a junction of channels, one veering to the north, the other towards the south. The former leads to the Canadian city of Cornwall, parts of which we can see. If we followed it, we would come to the entrance of the old Cornwall Canal and so have a splendid view of the downstream face of the power house. Canadians who know the river regret that one of the essential compromises in the final detailed agreement was that the two locks necessary to get up into Lake St Lawrence had to be built on the US side of the border, instead of as an integral part of the northern wing dam from the power house, on the line of the old Cornwall Canal and leading directly into the new lake. Visitors to the Canadian power house can see the special section of the concrete dam on the line of the old canal. When the time comes to duplicate the two locks in this section of the Seaway, there is little doubt but that they will be constructed in this location.

We must take the channel veering to the south, however, and make three sharp turns before we approach the Snell (previously Grass River) lock. We cannot but notice the swift current in this short reach, coming in from the north just before the lock is reached and deflected from the main channel by rock dykes that have clearly been carefully designed. Several model studies were made to ensure the safest and best channel arrangement in this critical location. High tension transmission lines will show us the direction in which the power house lies, hidden by wooded hills in the park. The importance of the location is indicated by a suspension bridge carrying a main road south from Cornwall, and by large industrial plants located here to take full advantage of the new power supply. Our pilot leaves the ship at this lock, his place taken by one of the next pilotage group who will guide the ship to the entrance of Lake Ontario. The locks are operated identically with the previous Canadian locks (except that the lock staffs have small battery-powered runabouts to facilitate their movements!), again excellently maintained, there being clearly a fine esprit-de-corps at all the locks. There follows a two-mile straight channel to the Eisenhower lock, this being a section of the Seaway in which several vessels can usually be seen at the same time. A tunnel can be identified beneath this second lock, conveying a road which leads to the Moses power station and State Park. The parking lot which adjoins the lock and the grandstand gives clear indication of the popularity of the Eisenhower lock with tourists, at least a few of whom will be seen at almost any hour of the day watching with fascination the passage of vessels through the great concrete structure.

Once the lock is passed, the channel remains straight, being still artificially excavated. If we are lucky, we may catch a fleeting glimpse of the Long Sault over-flow dam by looking astern but there is much to win our attention on the port bow. We sail slowly past the large rock-faced earthfill dykes that had to be constructed here in order to give the necessary water level at the power house, the elevation of the land behind them being clearly well below water level. We shall see a large aluminium works and much of the town of Massena, well protected in this way, but the necessity for the control dam at Iroquois will now be the more obvious. Once past the dyked area, we follow the channel located in the old river bed, although much deepened by dredging. Very gradually Lake St Lawrence narrows and we see that, just as with 'Lakes' St Louis and St Francis, it is just an enlargement of the river St Lawrence. This is dramatically shown when the floodgates which occupy the entire crest of the Iroquois Dam come into view, the dam appearing to be a small low structure. This is due to the water level being normally little different on the two sides but the dam structure, of mass concrete, is actually 67 ft high, founded on solid bedrock, now hidden beneath the impounded water. It is difficult to see the approach to the lock but we come to it along the Canadian shore, making a turn to port just before it is reached.

Sometimes (as when I passed through) the lock has the

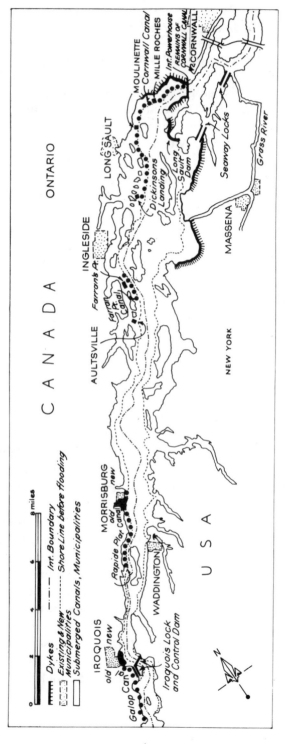

Map L: The St Lawrence: Cornwall to Iroquois

same water level at both ends; all gates are then opened for the transit of vessels. It requires but a glance, however, to see the special precautions that have been incorporated in the design of this part of the control structure. Instead of the usual single-gate mitre type closures, there are two pairs of steel sector gates (which move on sets of wheels on fixed tracks) at each end of the lock. One reason for the use of this type of gate here was to permit operation of the lock before the power pool was flooded when there was a drop in level of 18 ft, but they are also an emergency provision in case there was ever an accident downstream, since they are designed to withstand a head of 40 ft. There are the usual wire cable fenders at each end of the lock and also a full set of steel truss stop-logs stacked close to a stiffleg derrick at the downstream end. The operation of this lock and the Iroquois control dam are naturally important operations, the prime function of the dam being to control the water level of Lake Ontario, a matter of great public interest, in view of the development all around its shores, and so a matter of vital concern to the International Joint Commission.

Another suspension bridge comes into view between Prescott and Ogdensburg; this one spans the whole river and not just one channel as did that at Cornwall-Massena. This bridge is located close to the halfway mark of our sail from the last lock to Lake Ontario all of it through relatively flat but interesting country on both banks. We are still 68 miles from Lake Ontario, however, although now sailing in water that is almost at lake level. For much of this distance, the sail is through the world-renowned Thousand Islands section. There are indeed a thousand islands if not more, small and large, most of them still wooded even though now so widely used for ubiquitous summer residences. No words can really describe the beauty of this part of the river but it is still a busy waterway, requiring constant vigilance from the navigating officers and the pilot who finally leaves the ship by tender at Tibbett's Point as we enter Lake Ontario. Prior to this, however, we have sailed under yet another suspension bridge, this being one of two such spans that are the main parts of a multi-span river crossing through the islands known as the Thousand Islands or Ivy Lea Bridge, another important road connection between the two countries. At Tibbett's Point we shall be 190 miles upstream of Montreal, the passage usually taking westbound vessels about 22 hours and two or three hours less for those sailing downstream.

Now follows a sail of 160 miles across Lake Ontario, unless our vessel has to dock at Rochester, Toronto or Hamilton to unload cargo. On a foggy morning during this eleven-hour sail, with a choppy 'sea' and no land in sight, the sensation is exactly the same as being at sea, apart only from the absence of the tang of salt in the air. Approach to the Welland Canal, however, quickly dispels any thoughts of the sea. Near the protected harbour entrance of Port Weller we look ahead to the great escarpment stretching from east to west as an unbroken barrier and we, like all first visitors to this spot for a century and a half, wonder how our great vessel can ever mount such an obstacle. But we know from our study of the canal how the impossible is done.

We shall stop to pick up our next pilot from a small launch that comes out to meet us in the lake from the harbour of Port Weller, which is 7½ miles west of the mouth of the Niagara River. Its entrance is protected by two breakwaters 7,500 ft long, each 250 ft wide on the top, so that they are substantial structures. They were constructed using excavation from the canal, their ends properly equipped with the necessary guiding lights and navigation aids. Once inside the breakwaters, one is struck by the fine growth of trees on each of the embankments. This is the first evidence of the enlightened policy followed as soon as the canal was completed—of developing an extensive programme of reforestation along the canal banks, at all possible locations, to provide windbreaks as an aid to navigation in the relatively narrow channel for the large vessels using it, especially when passing through in ballast only with much superstructure exposed to prevailing winds. A small basin for maintenance vessels is passed on the east side of this entrance channel, a small landing stage for the pilot tender and a special landing for pleasure craft, just as there has been on each side of every lock. Here we are so close to it that we can readily read the notice advising those in charge of such small boats to use the adjacent telephone to consult with the lockmaster as to the movements they should make, a sensible and conveniently safe arrangement. On our starboard bow we see a long mooring wharf just before a slight turn brings us to the lower gate of Lock No 1. This lock is 865 ft long but otherwise identical with the next six, its normal lift being 46 ft although this may vary with changes in the level of Lake Ontario.

Locking through will at first appear to be just the same as the procedure in the Seaway locks proper but if we time our rise in the lock we shall see that it is not quite as fast as previously. This alone will remind us that we are now in a ship canal that was completed in 1931, having been in steady service ever since. One does not have to be an engineer in order to admire the forward-looking designs of those responsible for it half a century ago, and the excellence of the construction. There are a few more marks on the concrete corners of the approaches to the locks than previously but from every point of view the Welland Canal is today a fitting part of the new Seaway. On leaving the lock it is something of a surprise to see a large dry dock and shipbuilding works on the east (port) side, this being a private operation of the Port Weller Dry Docks Company. On the west (starboard) side, the canal is seen to be right in the city of St Catharines, residents of houses in pleasant streets close to the canal having an unrivalled view of the traffic it carries. There are, therefore, important bridges across the canal, starting at St Catharines, the first of the original twenty being a highway bascule bridge. There is now one high level fixed bridge but most of the important bridges are vertical lift bridges, for roads and railways, with a clearance beneath them of 120 ft. Lock No 2 comes at mileage 3.70, still in the city of St

Catharines, which has its own wharf 1½ miles farther on, close to a turning basin which can accommodate vessels up to 350 ft long. More bridges, and then Lock No 3 at mile 6.35, with more bridges close at hand and the twin locks up the escarpment clearly in view just a mile ahead. The first three locks, and Nos 7 and 8, are all single locks, but the twin locks are clearly seen as Lock No 4 is approached, the east side of the three twin locks being used by downward bound vessels, the west side by those upward bound for Lake Erie. A glance over the starboard bow will show a modern looking low building in a pleasant garden-like setting. This is the Control Building for the operation of the Welland Canal, a masterpiece of modern communication design. Connections from closed circuit television cameras at every lock, and constant radio contact, permit the controllers to watch the movement of every vessel in the canal system, an innovation in waterway control that has greatly speeded up transit through the Welland. How valuable this is will be well shown if, as is almost certain, we meet one or two of the '730s' on their way downstream, loaded to the last permissible inch of draft, their holds full of wheat. These vessels have been designed to fit into the Seaway locks with minimum possible clearance, their 730 ft length the maximum permitted, their beam of 75 ft only five feet less than the width of the locks—and it looks like five inches when one watches one of these giants slowly fill up a lock chamber. There were only four vessels of this size in 1959; now there are well over fifty. This explains why the number of transits of the Seaway by individual vessels has steadily decreased since its opening, even though the tonnage it carries has increased since then by about two and one half times! Generally single screw vessels, the 730s are manoeuvred by their masters with great skill, their handling assisted by special cross tunnel mounted engines near their bows, permitting a quick 'push' to either side when so actuated from the bridge. Bridges are usually in special superstructures near the bow but quite a few vessels have the same sort of arrangement near the stern, a mystery of naval architecture that I have not yet been able to solve.

The steady rise through the three locks is an operation that never fails to impress; even the most hardened of canal travellers still watches it with respect and with admiration for the superb engineering design that makes it possible. Lines are worked off central 'islands' between the pairs of locks. With vessels often in each, one ascending while the other descends, the lock staffs are kept busy indeed. Even the best of photographs fails to do justice to the majestic sight of these six locks as one sails into the first and starts the great rise of 140 ft but some impression of what they look like can be gained from page 139. On the west side one sees a small water power station that supplies all the canal works with their electrical supply. On the east is a chain of ponds and weirs for conveying waste water down to the canal below; a close look will show signs of the locks of the third canal that were here used for this purpose. Once in Lock No 4, a vessel proceeds straight up through No 5 and No 6 into a half mile stretch beyond this, in which traffic again comes into the one channel for passage through Lock No 7 which is again a single lock. Vessels have sometimes to wait in Lock No 6 if there is a vessel ahead waiting to enter Lock No 7. But this upper lock is a good place for stopping, since on a clear day the view back over the lower part of the canal and Lake Ontario is awe-inspiring. All too soon we shall be moving out of Lock No 6, however, into the short approach channel to single Lock No 7 which will take us up to the level of Lake Huron.

Here is where one realises best of all, perhaps, the intensity of traffic through the canal and the almost obvious necessity of twinning all the locks within the foreseeable future. This is naturally being studied by the Canadian Seaway Authority, now that the bypass works are complete. Construction of a completely new canal, but on the US side of the Niagara River, has also been studied by the US Army Corps of Engineers. Their report of 1973 showed that the cost would be $2.6 billion, an expenditure that is said to be not economically justified 'based solely on US transportation benefits'. An international study of possible improvements of the Seaway system as a whole is recommended and this may be confidently anticipated, despite the nationalist voices still to be heard occasionally from the USA objecting to the fact that the Welland Canal is located wholly in Canada, even though serving so well the needs of both partner countries.

Such talk of enlargements may seem strange in view of the opening of the Seaway as recently as 1959. Traffic increases, however, provide the answer. In 1973, for example, the tonnage passing through the Sault Ste Marie Locks was 109.7 million tons; through the Welland Canal 67.2 million; and through the international section 57 million, all tonnages being appreciably greater than those carried in 1972. Iron ore accounts for about two-thirds of the tonnage at the Soo, wheat about one quarter, miscellaneous freight making up the remainder. Tolls have to be paid on all cargoes but the rates are still those fixed at the Seaway opening in 1959, all efforts to increase them since that time having been unsuccessful. The result of world-wide inflation as it has affected North America has resulted in a serious financial situation. In the case of the Canadian Authority, capital debt has increased to $775 million since toll income has not been sufficient to pay interest charges on capital. Results for 1973 led to the first annual operating deficit, an amount of about one million dollars. This is a situation that is also naturally under study, one that is most complex in view of public attitudes to 'subsidies' for most forms of transport.

As we continue our slow sail up the Welland Canal, however, we can leave these weighty problems to Seaway management and, ultimately, the two governments. Once past Lock No 7 we are in a widening section of the canal with a great paper mill straight ahead, well equipped with its own wharves and special turning basin, a wharf running continuously along the west bank of the canal until a turn to the west which is suddenly seen just before the Ontario Paper Company's plant is reached. At the turn, the canal narrows to a guard structure not unlike one end of a lock, complete with one set of guard gates, this being one of the many safety and precautionary measures with which the canal is equipped—some seen, some unseen, but all designed to take care of any possible eventuality or accident. The principal purpose of this particular feature is to maintain the water level in the long upper level of the canal in case anything untoward ever happened to Lock No 7. Then follows a clear sail of 15½ miles to the guard lock at Port Colborne, through well developed country, passing numerous industrial plants. Originally the canal passed right through the centre of the town of Welland, with the inevitable plethora of bridges, at about mileage 19 from Port Weller.

Not only has very careful maintenance work been well done down through the years, especially on the large lock gates and their bearings, but successive improvements have been made to the original layout of the canal. For the most part these were relatively small localised operations, although one significant straightening operation was carried out at Port Colborne in the late thirties. A main objective of all improvements was to reduce transit time. Starting in 1967, however, a major improvement was effected in the vicinity of the town of Welland by the construction of an entirely new, straight, channel with a length of 8.3 miles. This new section bypassed the old winding section through Welland completely and so eliminated six bridges, one of which (with a central pier) had long been regarded as a real problem especially by the masters of the larger vessels using the canal. In place of the bridges, two major tunnels were constructed 'in the dry' before completion of the canal excavation. Roads and railways throughout the district were re-routed so that all could use the two tunnels, one conveying both rail tracks and roads, and the other roads only. The construction of these facilities, at a cost of $40 million and $13 million respectively, in excavations more than 90 ft deep, has provided Canadian civil engineering practice with

some invaluable experience. Almost 100 miles of new railway track were necessary and this alone cost about $50
million.[2]

It is small wonder, therefore, that the new bypass canal
cost $188 million by the time it was complete, excavation of
the 350 ft wide channel costing only $40 million of the total.
Use of the new bypass is estimated to cut out about half an
hour from the usual 13 to 14 hour passage of the entire canal,
and this will increase the capacity of the Welland Canal by
from 3 to 4 per cent. The expenditure of such a great sum for
such an apparently small improvement in operation might at
first sight seem paradoxical but it will be seen to be well
justified when we come to see what has been happening to
traffic on this vital waterway. Merely as an indication of the
need for the improvement is the fact that in the 1973 operating season, traffic increased by 4.7 per cent over the previous
record tonnage of the year before. It is clear that the Welland
Bypass, as it is already known colloquially, will prove to be
but one more step in the steady development of the Welland
Canal.

The bypass was ready for use at the opening of the 1973
season of navigation but advantage was taken of the holding
in Canada of an important international meeting to have
another grand ceremony to mark its official opening. The
meeting was one of the regular meetings of the Permanent
International Association of Navigation Congresses which
convened in Ottawa. Special arrangements were made for
the delegates, and many important figures involved in both
Canadian and US shipping, to gather at the Welland Canal
on 14 August 1973. Some special guests made a ceremonial
journey through the bypass on the *Taranteau*, one of the
largest of the new upper lakers of Canada Steamship Lines,
the formal ceremony of dedication taking place at the new
Welland wharf. (Refreshments included Canadian 'cham-
pagne' from five nearby vintners, the canal being located
immediately adjacent to the wine-growing area of the Niag-

ara peninsula, a feature of the Canadian scene but little known outside its borders.)

The guard lock at Port Colborne (No 8), which is reached soon after the new bypass has been left on the sail towards Lake Erie, is 1,380 ft long and 80 ft wide as usual. It has been called 'the longest canal lock in the world' but its utility in guarding this southern entrance to the Welland Canal is of far more significance than its great length. It leads into the short final stretch of channel and so into the harbour which has been developed at Port Colborne. In a sense this is a double harbour since the old guard locks for the third canal are still there although now separated from the new ship channel and anchorage area. This has an area of 37 acres and is well equipped with service quays and other corresponding facilities. About 3½ miles of wharf are available (with varying depths of water) for winter berthage of those lake vessels that lie up for the winter at Port Colborne. The government grain elevator still dominates the sky line but the area around has developed into an important industrial complex, all tribute to the influence of the canal. We shall make a very brief stop since there will be another change of pilot here. Although the captain can navigate his vessel for most of the length of Lake Erie, the channels in the shallow western end, the approach to the Detroit River, the passage through Lake St Clair and then up the St Clair River into Lake Huron all necessitate the expert assistance of experienced pilots. Transit of the lake, even though it is the shallowest of all the Great Lakes, is again akin to ocean travel until approach is made to one or other of the lake ports such as Cleveland or Toledo. This is not exactly canal travel and so we must quickly complete our long journey that owes so much to master canals—enjoying the active scenes of industrial development along the US side of the Detroit River in contrast to the generally rural scenes on the Canadian side, this situation being reversed as we proceed up the St Clair River and pass Canada's 'Chemical City' of Sarnia with its growing complex of important chemical plants.

Sailing north up Lake Huron is the same as being at sea again. It took us well over thirteen hours to transit the Welland Canal and another twelve hours to sail through Lake Erie. Passage between Lake Erie and Lake Huron is inevitably at rather slower speed, but if weather permits, full speed will be the order of the day hereafter, apart from our passage through the St Mary's River, reached after a sail of 223 miles. We could divert before reaching St Joseph Island in order to sail through US waters past the historic Michimilimackinac Island, under another great suspension bridge and so into Lake Michigan on our way to Chicago, Milwaukee or one of the other ports on this important all-US Great Lake. If our vessel were small enough we could transit the Chicago Drainage Canal and so reach the headwaters of the Mississippi River down which we could sail to the Gulf of Mexico, such are the conveniences of these waterway connections. Let us, however, continue along the international route, up the lovely St Mary's River with its sharp but readily navigated bends, through one of the five locks at Sault Ste Marie and so into Lake Superior, greatest and deepest of all the Great Lakes. We could sail no less than 383 miles westward up this inland sea—for such it is—if we were bound for Duluth, the US port at the head of the lake. Let us rather terminate our journey at the Canadian lakehead, the joint harbours of Port Arthur and Fort William, two communities now happily united into the one city of Thunder Bay.

The well protected harbour is magnificently situated, dominated by one of the continent's greatest concentrations of grain elevators in which wheat from the prairies, delivered by rail, is stored for eventual transfer into either ocean-going vessels or lakers which will carry it all the way to Montreal or even to St Lawrence ports further downstream. Here also are loading facilities for iron ore coming from Canada's Steep Rock mine. A wide variety of harbour services make the waterfront a busy place indeed. If we are fortunate enough to see a loaded foreign vessel sail out of the harbour on its way down the 2,300 miles to the Atlantic and so to Europe, we can recall the transits it will make of the Welland Canal and the Seaway Locks and so reflect that, despite all the changes through the years, there is still potent meaning when one talks of the Canals of Canada.

CHAPTER 15

L'envoi

The St Lawrence Seaway has already seen on its waters many splendid ships but nothing quite like the flotilla that sailed up the international section at the end of June 1959. It had been opened unofficially at the start of the navigation season earlier in the year, the first vessel to sail into the St Lambert lock being the C G S *d'Iberville* (one of the largest Canadian icebreakers) carrying a large official party. On 26 June, however, H M Y *Britannia* sailed into this first lock after passing through a specially erected ceremonial gateway, carrying Her Majesty Queen Elizabeth II of Canada and President Eisenhower of the United States officially to declare the Seaway open. This they did with singularly appropriate words in the presence of a distinguished company. *Britannia* and her escorts then proceeded through the Côte Ste Catherine lock into Lake St Louis where a naval review took place.

The royal yacht then proceeded up the Seaway but dense fog caused some delay so that Her Majesty had to complete her journey to the Eisenhower Lock by road. Here she was received by the Vice-President of the United States and a distinguished company of US citizens who had driven over to the great power house, where a special platform had been erected exactly on the international boundary at the downstream side of the building. The historic ceremony culminated in the joint unveiling of a special plaque, built into the power house structure, by Her Majesty the Queen and the Vice-President.

The plaque stands there today as the only indication that this is the international boundary. Here are no guards, no barbed wire but just the low hum of the nearby generators, using the fall of water in the great river to develop power 'for the use and convenience of man'. Unfortunately, but quite understandably, this location cannot be visited by the public. Since it lies between the Canadian and US sections of the power house, and can be seen only by special permission from one or other of the power authorities, this memorial is therefore but little known to citizens of either country.

The plaque is a polished tablet of black stone about fifteen feet square. It carries beautifully prepared metallic reproductions of the coat of arms of the Dominion of Canada and the Great Seal of the United States of America. Between these two symbols are engraved some simple words, set deep in the stone but shining out as a golden message, one that all who see them want to share. They may well serve to bring this volume to a close since these are the words that Her Majesty read as the curtain was pulled apart:

THIS STONE BEARS WITNESS TO THE COMMON PURPOSE OF TWO NATIONS WHOSE FRONTIERS ARE THE FRONTIERS OF FRIENDSHIP, WHOSE WAYS ARE THE WAYS OF FREEDOM, AND WHOSE WORKS ARE THE WORKS OF PEACE.

Notes on the Text

Chapter One

1. *The Fur Trade in Canada,* H.A. Innis, Toronto, 1962, p. 392.
2. *Fur Trade Canoe Routes of Canada; Then and Now,* E. W. Morse, Ottawa, 1968, p. 20.
3. *Montreal: Island city of the St Lawrence,* Kathleen Jenkins, New York, 1966, p. 234.
4. *American Notes for General Circulation* (1842), Charles Dickens, London, 1972, p. 248.
5. 'Steam Navigation on the Ottawa River,' H. R. Morgan, *Papers and Records,* Ont. Hist. Society, vol. 23 1926, pp. 370-383.
6. *Connecticut River,* Marguerite Allis, New York, 1939, p. 127.
7. *The Erie Canal,* an American Heritage book (good introduction), New York, 1964, pp. 153. See also *Erie Water West: A History of the Erie Canal, 1792-1854,* Ronald E. Shaw, Lexington, 1966.
8. 'New York State Barge Canal System,' E.C. Hudowalski, *Proceedings of the American Society of Civil Engineers,* 1959, paper 2176 (WW2).
9. *General Report of the Commissioners of Public Works for the year ending 30 June 1867,* J. C. Chapais, Ottawa, 671 pp.

Chapter Two

1. *Transportation and Communication in Nova Scotia,* R. D. Evans, 1936, unpublished thesis in Nova Scotia Archives, Halifax.

2. Quoted in Crawley H. O., 'Proposed Canal to unite the Bay of Fundy with the Gulf of St Lawrence,' *Professional Papers of the Corps of Royal Engineers*, VIII pp. 186-193, 1846.
3. 'The Chignecto Ship Railway,' S. T. Spicer, *Canadian Geographical Journal*, May 1961, pp. 178-183.
4. *St Peter's History, Municipal Project of Richmond County* (Notes for the Nicholas Denys Museum, n.d., 4 pp.)
5. See (1) pp. 33-67.
6. *Canals of Canada* (1953), p. 28.
7. Extracts from (typed) Histories of H M C S Destroyers *Patriot* and *Champlain*, Dept. of National Defence, Ottawa.
8. See (1) pp. 33-67; and also 'Railway and Frost ended Canal Use in Dartmouth,' D. Grant, Halifax *Mail-Star*, 14 August 1961.
9. 'The Shubenacadie Canal,' H. E. Chapman, *The Atlantic Advocate*, April 1965, pp. 45-48.
10. 'The Shubenacadie Canal,' E. H. Keating, *Transactions of the American Society of Civil Engineers, 12*, 1883, pp. 436-440.
11. *Canals of Canada*, 1953, Ottawa, p. 8.
12. 'The Strait of Canso,' *The Atlantic Advocate*, June 1973, pp. 15,16.

Chapter Three
1. *Canals of Canada* (1953), p. 28.
2. *Canals of Canada* (1953), p. 29.
3. 'New York State Barge Canal System,' E.C. Hudowalski, *Proc. Am. Soc. of Civil Engs.*, Sept. 1959, Paper 2176 (WW 2); see also various leaflets of the New York State Department of Transportation.
4. *American Notes for General Consumption* (1842), C. Dickens, London, 1972, p. 254.
5. 'Richelieu Project Revived,' *The Gazette* (Montreal), 10 March 1962.

6. 'The Richelieu, Historic Waterway of Eastern Canada,' W. E. Greening, *Can. Geog. Journ.*, March 1958, Vol. 56, pp. 84-93.

7. 'Lake Champlain Dam Work Gets Commission Approval,' *The Gazette* (Montreal), 11 Jan. 1937.

8. 'Mark Anniversary of Chambly Canal,' *The Gazette* (Montreal), 6 September 1943; see also *Canals of Canada* (1953) p. 29.

9. 'The Richelieu Waterway,' Guy Tombs, *The Gazette* (Montreal), 19 March 1958.

10. *Report of the Royal Commission on Government Organization,* Ottawa, 1963, vol. 3. p. 370

See also the well-illustrated booklet *The Chambly Canal/ Canal de Chambly* published in 1943 in St. Jean, Que., by the Centenary Committee. A copy of this rare 90-page brochure was kindly given to me by Mr L. Tombs when this book was being prepared.

Chapter Four

1. 'The Ottawa River Canals and Portage Railways,' R. F. Legget, *Transactions of the Newcomen Society, 40,* 1967-68, pp. 61-73.

2. *Canals of Canada* (1963), Ottawa, p. 12.

3. *Rideau Waterway,* R. F. Legget, Toronto, 1955, 249 pp.

Chapter Five

1. 'The Trent-Severn Waterway in Ontario,' Enid S. Mallory, *Can. Geog. Jour.*, May 1963, Vol. 66, pp. 140-153.

2. *Journals of Anne Langton,* ed. by H. H. Langton, Toronto, 1950, p. 47.

3. 'Seaway Recalls Failure of Trent Canal,' W. A. Collins, *Globe and Mail,* Toronto, 18 Sept. 1954; see also (1).

4. *Canals of Canada,* (1953), p. 38.

See also Waterways to Explore - The Trent, Mary Ainslie, Toronto, Ont. Dept. of Travel and Publicity, 1946, 22 pp.

Chapter Six

1. *Rideau Waterway,* R. F. Legget, Toronto, 1955; see pp. 139-142.
2. *Canals of Canada* (1953), p. 40.
3. From a personal communication of G. B. Williams, Senior Asst. Deputy Minister, Dept. of Public Works, Canada, 17 August 1973.
4. Information supplied by S. B. Panting, Director, Engineering Services Branch, (Ontario) Ministry of Natural Resources, Toronto, 19 February 1973.
5. See (3).
6. *Fifteen Years Sport and Life in the Hunting Grounds of Western America and British Columbia,* London, 1900; see also note by W. E. Trout in *Bulletin No. 7* of the American Canal Society, November 1973, p. 8.
7. Information kindly supplied by Mrs. Kersten Mueller, Librarian, Fort Frances Public Library; see also Final Report of the International Joint Commission on the Regulation of Lake of the Woods, 1917 Ottawa, pp. 198-203.
8. *The Canals of Canada,* W. Kingsford, Toronto, 1865, see pp. 79-80.
9. Ibid, pp. 80-81.
10. *Kingston 300, a Social Snapshot,* by Kingstonians, Kingston, 1973, see p. 132.
11. *History of the Town of Newmarket,* Newmarket, n.d., see p. 270.

Chapter Seven

1. 'Georgian Bay Ship Canal: Report upon Survey, with Plans and Estimates of Cost,' xxii, 599 pp., in *Sessional Papers,* vol. 10, First session of the eleventh Parliament of the Dominion of Canada, 1909.
2. 'The Georgian Bay Canal,' R. J. Morgan, *Can. Geog. Jour.,* Vol. 78, March 1969, pp. 90-97.
3. *The Upper Ottawa Valley,* C. C. Kennedy, Pembroke, 1970, see p. 138-139.

Chapter Eight

1. General introductions are given by: *The St Lawrence River*, H. Beston, New York, 1942; and *The St Lawrence Valley*, K. Lefolii, Toronto, 1970.

Chapter Nine

1. *A Study of the St Lawrence Waterway with particular reference to the Lachine Section*, A. J. Grant, privately printed, June 1952.
2. 'Lachine Canal; its Part in Canada's History,' A. F. Cross, Montreal *Star*, 27 July 1935.
3. *Canals of Canada* (1953), pp. 10-12.

Chapter Ten

1. *Canals of Canada* (1953), pp. 12-13.
2. Ibid., pp. 13-14.
3. Ibid., pp. 14-16.
4. Ibid., pp. 15-17.
5. 'Final Voyage along a great Waterway,' R. Rice, a *Canadian Press* report, 12 December 1958.

Chapter Eleven

1. *The Welland Canal Company*, H. J. Aitken, Cambridge Mass., 1954, 178 pp.
2. *Biography of the Hon W. H. Merritt M.P.*, J. B. Merritt, St Catharines, 1887, 429 pp.
3. *The Canals of Canada: a Prize Essay*, T. C. Keefer, Montreal, 1850, see pp. 63,64. (A copy of this rare 111-page pamphlet was kindly given to me by Dr P. Harker of Ottawa when this book was being written.)
4. *The Canadian Canals*, W. Kingsford, Toronto, 1865, p. 61
5. *Report of the Select Committee of the Upper Canada House of Assembly on the Welland Canal Company*, Toronto, 1836, 575 pp.
6. *Canals of Canada* (1953), p. 18
7. Ibid., p. 18.

8. 'The Welland Canal,' W. A. O'Neil, *Proc. Am. Soc. of Civil Engs.*, March 1958, Paper 1570 (WW2).

Chapter Twelve

1. *Travels through the Canadas*, G. Heriot, London, 1807, p. 199.
2. *Annual Report of the Superintendent and Collector of the St Mary's Falls Ship Canal for the year 1878*, Lansing Mich., 1879; this Report has a 12-page supplement giving a detailed history of the US locks at Sault Ste Marie from 1836 to 1878 (my copy being another gift from Dr P. Harker).
3. *Canals of Canada* (1953), p. 27.

Chapter Thirteen

1. *The Canals of Canada; a Prize Essay*, T. C. Keefer, Montreal, 1850, p. 35.
2. Quoted in (4) p. 49.
3. *The St Lawrence Deep Waterway: A Canadian Appraisal*, C. P. Wright, Toronto, 1935, 450 pp. (a detailed and authoritative review of the complex history of the Seaway until 1935).
4. *The St Lawrence Seaway*, T. L. Hills, London, 1959, 157 pp. (one of the several books published at the time of the opening of the Seaway, a concise well illustrated review).
5. 'The St Lawrence Seaway, 27-ft Canals and Channels,' W. Grothaus and D. M. Ripley, *Proc. Am. Soc. of Civil Engs.*, January 1958, Paper 1518 (WW2); and 'Canadian Section of the St Lawrence Seaway,' L. H. Burpee, Paper No. 2420 (WW1) in same *Proceedings*, March 1960.

Chapter Fourteen

1. 'Canadian Section of the St Lawrence Seaway,' L. H. Burpee, *Proc. Am. Soc. of Civil Engs.*, March 1960, Paper No. 2420 (WW1), see p. 51.

2. 'Re-routing the historic Welland Canal,' G. V. Sainsbury, *Can. Geog. Jour.*, September 1974, Vol. 88, pp. 36-43.

The convenience of British readers interested in the more technical aspects of the development of Canadian canals may be served by this list of articles published down the years in the pages of the British journal *Engineering*, kindly provided by Mr E. S. Turner of Ottawa:

	Vol.	Page	Year
Proposed Welland Canal	4	505	1867
Huron and Ontario Ship Canal	5	33	1868
Gates: Welland and St Lawrence Canals	9	130	1870
Sault Ste Marie Canals	41	357	1886
Chignecto Ship Railway	48	408	1889
Georgian Bay Ship Canal	67	134	1899
Lock on Trent Valley Canal at Peterborough	67	622	1899
Soulanges Canal	68	536	1899
Trent Valley Lift Lock	81	333	1906
Georgian Bay Ship Canal	89	638	1910
Peterborough Lift Lock	101	548	1916

These references were not consulted in the preparation of this book.

APPENDIX 1

Some Canadian Canal Statistics

ST PETER'S

Year Opened 1869

Year	Tonnage Vessels	Freight	Number of Vessels
1875	30,581	18,116	807
1880	7,232	4,045	153
1885	68,716	20,160	1,148
1890	70,985	32,231	1,294
1895	16,416	9,828	248
1900	115,783	73,813	1,628
1905	104,959	81,077	1,595
1910	107,053	85,951	1,470
1914	80,665	54,180	1,200
1918	69,287	59,716	1,071
1925	81,363	35,691	1,196
1930	79,784	59,973	865
1935	83,163	54,592	1,021
1939	104,897	79,015	1,048
1945	55,578	21,665	734
1950	30,005	7,805	669
1955	27,324	6,783	784
1960	15,831	723	555

Source: Statistics Canada   # Not reported

Appendix 1

CHAMBLY AND ST OURS LOCK

Year Opened 1849

Year	Tonnage Vessels	Freight	Number of Vessels
1851	90,691	110,727	1,727
1856	151,070	129,666	2,617
1861	122,694	116,239	2,119
1867	418,644	410,430	5,228
1870	543,220	518,334	6,611
1875	261,344	242,115	2,925
1880	327,176	202,067	3,296
1885	230,721	184,212	2,210
1890	231,747	202,407	2,108
1895	331,836	359,021	3,262
1900	300,755	348,561	2,841
1905	379,112	447,069	3,343
1910	467,246	669,299	4,219
1914	294,408	436,900	2,694
1918	263,722	369,186	2,297
1925	183,541	203,720	1,476
1930	81,989	99,998	779
1935	48,417	44,219	468
1939	107,815	111,677	806
1945	48,190	46,578	296
1950	104,526	105,373	692
1955	100,008	97,130	701
1960	124,027	106,699	796
1965	#	87,741	#
1970	#	12,797	#

#Not reported   Source: Statistics Canada

OTTAWA RIVER CANALS

Year Opened   1833-1843

Year	Tonnage Vessels	Freight	Number of Vessels
1851	* 102,407	105,933	1,987
1856	* 177,686	169,401	2,874
1861	* 219,675	199,097	3,486
1867	* 450,942	343,139	6,676
1870	* 554,879	488,228	7,356
1875	334,798	497,494	4,801
1880	541,478	644,549	5,202
1885	457,535	763,236	3,572
1890	400,239	651,355	2,829
1895	289,116	541,220	2,195
1900	270,116	389,145	2,114
1905	257,897	390,771	2,152
1910	452,820	385,261	2,601
1914	419,773	335,132	2,472
1918	253,490	167,170	1,488
1925	317,855	214,190	2,246
1930	693,000	540,933	3,258
1935	343,937	289,526	1,802
1939	346,225	301,671	1,676
1945	291,187	258,172	1,256
1950	340,084	294,604	1,117
1955	235,654	206,525	952
1960	341,475	278,200	1,350
1965	#	615	#
1970	#	#	#

Not Reported   * Ste Anne's Lock only   Source: Statistics Canada

Appendix 1

RIDEAU CANAL

Year Opened 1832

Year	Tonnage Vessels	Freight	Number of Vessels
1851	#	#	#
1856	#	#	#
1861	+334,157	213,491	4,968
1867	+574,614	470,242	9,014
1870	+672,674	626,714	9,700
1875	185,833	163,382	2,913
1880	150,683	101,298	2,682
1885	120,493	87,944	1,910
1890	140,678	113,574	2,238
1895	168,852	88,753	2,375
1900	191,515	75,432	2,579
1905	186,559	59,864	4,715
1910	183,242	134,881	2,815
1914	179,575	151,739	2,635
1918	78,803	54,136	1,164
1925	105,036	85,785	1,496
1930	51,452	28,210	517
1935	42,298	20,426	624
1939	5,478	2,009	258
1945	3,297	451	65
1950	4,456	1,215	135
1955	4,488	413	138
1960	3,126	72	94
1965	#	#	#
1970	#	#	#

\# Not Reported   Source: Statistics Canada
+Includes Ottawa River Canals but not Ste Anne's Lock

MURRAY CANAL

Year Opened 1889

Year	Tonnage Vessels	Freight	Number of Vessels
1890	101,504	18,783	865
1895	162,414	11,324	568
1900	213,179	19,067	745
1905	228,837	29,421	707
1910	379,450	177,941	1,308
1914	213,636	83,907	971
1918	115,719	44,735	453
1925	46,703	1,174	493
1930	69,700	2,316	443
1935	39,236	4,921	295
1939	21,120	3,707	653
1945	28,795	1,915	95
1950	33,988	2,969	66
1955	21,330	667	53
1960	19,183	130	194
1965	#	#	#
1970	#	#	30

#Not Reported   Source: Statistics Canada

CANSO LOCK

Year Opened 1955

Year	Tonnage Vessels	Freight	Number of Vessels
1955	31,104	13,199	495
1960	764,446	803,376	1,228
1965	1,089,046	1,082,405	1,458
1970	1,381,232	1,469,415	1,473

Source: Statistics Canada

Appendix 1

TRENT CANAL

Year Opened 1844

Year	Tonnage Vessels	Freight	Number of Vessels
1851	#	#	#
1856	#	#	#
1861	#	#	#
1867	#	#	#
1870	#	#	#
1875	#	#	#
1880	430	18,224	8
1885	3,880	25,707	79
1890	51,800	24,679	1,304
1895	85,315	32,266	1,947
1900	100,970	43,572	2,212
1905	122,735	45,231	2,046
1910	172,085	46,263	3,442
1914	174,647	67,715	3,647
1918	172,133	64,893	3,549
1925	98,458	36,302	2,701
1930	54,848	23,875	1,726
1935	32,113	14,157	2,824
1939	72,908	28,985	3,719
1945	60,737	50,612	3,011
1950	1,175	602	76
1955	1,016	102	66
1960	19,183	130	194
1965	#	13	#
1970	#	53	#

#Not Reported   Source: Statistics Canada

BURLINGTON CANAL

Year Opened 1851

Year	Vessels	Tonnage Freight	Number of Vessels
1851	481,910	58,108	2,533
1856	450,043	97,104	885
1861	294,244	178,674	1,784
1867	282,718	172,384	1,661
1870	270,530	159,967	1,546
1875	266,645	125,524	1,253
1880	351,863	98,608	989
1885	115,089	73,174	425

ST ANDREW'S LOCK

Opened in 1910

Year	Vessels	Tonnage Freight	Number of Vessels
1910	44,887	8,283	202
1914	106,044	42,013	334
1918	28,062	4,640	130
1925	71,843	70,799	384
1930	115,645	89,715	847
1935	63,508	19,063	510
1939	72,595	20,526	616
1945	60,781	11,216	568
1950	52,123	14,901	416
1955	38,368	8,112	353
1960	16,467	4,099	301
1965	#	1,624	#
1970	#	1,791	#

#Not Reported   Source: Statistics Canada

Appendix 1

ST LAWRENCE CANALS

Year Opened   1842

| Year | Tonnage | | Number of Vessels |
	Vessels	Freight	
1851	526,210	450,400	6,934
1856	715,041	634,536	7,889
1861	1,009,469	886,908	11,042
1867	1,122,916	906,299	11,832
1870	1,354,102	1,215,067	13,572
1875	1,269,349	907,640	10,337
1880	2,015,582	1,072,556	11,340
1885	1,550,696	734,280	9,030
1890	1,741,447	853,853	10,498
1895	1,747,416	828,228	8,746
1900	2,138,357	1,309,066	9,658
1905	2,368,201	1,752,855	9,996
1910	3,392,539	2,760,752	10,226
1914	5,281,725	4,391,493	10,245
1918	3,810,025	3,031,134	7,297
1925	6,630,227	6,206,988	12,556
1930	5,971,060	6,179,023	9,519
1935	6,057,139	6,873,655	10,163
1939	6,484,320	8,340,165	9,359
1945	5,882,526	6,947,870	7,623
1950	8,233,472	9,969,271	10,147
1955	9,046,487	11,446,620	10,388
1960	+17,333,605	20,752,161	12,040
1965	+28,892,723	43,378,663	7,868
1970	+31,348,113	51,196,586	6,716

+The Seaway   Source: Statistics Canada

WELLAND CANAL

Year Opened   1829

Year	Tonnage Vessels	Freight	Number of Vessels
1851	772,623	691,627	5,693
1856	1,179,246	976,556	6,766
1861	1,327,672	1,020,483	6,708
1867	993,938	933,260	5,305
1870	1,357,117	1,319,290	6,740
1875	1,009,216	1,038,050	4,270
1880	882,276	819,934	4,104
1885	681,947	784,928	2,733
1890	1,122,469	1,016,165	2,888
1895	1,068,373	869,595	2,222
1900	1,012,812	719,360	2,399
1905	1,101,495	1,092,050	1,595
1910	2,148,517	2,326,290	2,544
1914	3,635,695	3,860,969	3,692
1918	2,595,389	2,174,298	3,192
1925	5,567,136	5,640,298	5,966
1930	5,574,567	6,087,910	5,352
1935	+8,207,552	8,953,383	5,092
1939	9,900,563	11,727,553	6,063
1945	11,855,532	12,962,332	6,291
1950	13,159,442	14,740,573	7,436
1955	19,059,617	20,893,572	9,334
1960	23,907,489	29,280,737	8,169
1965	38,547,495	53,436,596	8,352
1970	45,986,757	62,965,510	7,200

+Fourth Canal opened in 1932   Source: Statistics Canada

CANADIAN LOCK AT SAULT STE MARIE

Year Opened 1895

Year	Tonnage Vessels	Freight	Number of Vessels
1895	749,626	595,837	1,192
1900	2,194,748	2,035,677	3,081
1905	5,537,637	5,473,406	5,662
1910	23,361,198	36,395,687	7,972
1914	17,301,162	27,599,184	5,977
1918	10,030,542	12,913,711	5,059
1925	5,687,547	1,634,970	3,534
1930	3,481,576	1,691,471	2,959
1935	2,951,553	1,932,047	2,896
1939	4,200,359	2,775,769	3,327
1945	4,207,754	2,078,886	2,676
1950	3,544,194	2,301,763	3,666
1955	4,119,758	2,201,075	4,908
1960	3,525,251	1,720,622	4,855
1965	3,784,922	1,406,609	3,484
1970	2,582,928	1,352,016	3,099

Source: Statistics Canada

CANADIAN AND USA LOCKS AT SAULT STE MARIE

Year Opened   1855

Year	Tonnage Vessels	Tonnage Freight	Number of Vessels
1851		-	-
1856	101,458	33,817	290
1861	276,639	87,847	538
1867	556,899	325,357	1,305
1870	690,826	539,883	1,828
1875	1,259,534	833,465	2,033
1880	1,734,890	1,321,906	3,503
1885	3,035,937	3,256,628	5,380
1890	8,454,435	9,041,213	10,557
1895	16,806,781	15,062,580	17,956
1900	22,315,834	25,643,073	19,452
1905	36,617,699	44,270,680	21,679
1910	49,856,123	62,363,218	20,899
1914	41,986,339	55,369,934	18,717
1918	61,100,244	85,680,327	20,610
1925	69,237,687	81,871,699	20,651
1930	54,830,314	72,897,895	16,820
1935	41,567,383	48,292,973	12,959
1939	56,629,185	69,849,304	17,073
1945	86,932,685	113,227,316	22,492
1950	80,319,285	106,195,738	21,004
1955	91,732,406	114,553,735	21,941
1960	#	91,774,624	21,461
1965	#	95,593,250	17,476
1970	#	101,402,636	16,710

Source: Statistics Canada   # Not reported

Canals of Canada, 1974

Date originally opened	Length in miles	Number of locks	Lock dimensions in feet	Normal draft in feet	Lift in feet
		ST PETER'S CANAL			
1869	0.50	One *4 gates*	300 x 47	17	V
		CANSO CANAL			
1955	0.78	One *Sector gates*	820 x 80	28	V
		ST OURS CANAL			
1849	0.12	One	339 x 45	12	5
		CHAMBLY CANAL			
1843	11.78	Nine	120 x 23	6.5	80
		STE ANNE CANAL			
1843	0.12	One	200 x 45	9	V
		CARILLON CANAL			
1963	0.50	One	188 x 45	9	65
		RIDEAU CANAL			
1832	123.53	47	134 x 34	5.5	277 up 162 down
		MURRAY CANAL			
1889	5.15	None	124 at W L	9.5	
		TRENT CANAL			
1930 *in full*	240.56	44 *and marine rly*	varies	6-*see text*	598 up 260 down

Date originally opened	Length in miles	Number of locks	Lock dimensions in feet	Normal draft in feet	Lift in feet
		MUSKOKA LOCKS			
1873		One	83 x 12	8.5	4
1874		One	175 x 33	8.5	4
		HUNTSVILLE LOCK			
1874		One	88 x 20		8
		MAGNETAWAN LOCK			
		One	112 x 28	5	
		ST ANDREWS LOCK			
1916		One	215 x 46	10	V
		ST LAWRENCE SEAWAY			
1959	190	Seven	768 x 80	30	211
		WELLAND CANAL			
1932	27.60	Eight	859 x 80	30	327
		SAULT STE MARIE CANAL			
1895	1.38	One	900 x 60	18.25	19

Notes: Lock dimensions are the minimum in each canal; in some cases, dimensions include a few additional inches.

Both *drafts* and *lifts* are given, when possible, for normal conditions; the use of 'V' indicates regular variations which are described in the text.

Suggestions for Further Reading

A History of Transportation in Canada by G. P. deT. Glazebrook, Toronto, McClelland & Stewart, 1938, reissued as two vols., paperback, Toronto, 1964 provides an excellent general introduction to all aspects of transportation in Canada. Chapters 3 and 13 provide a concise summary of canal development up to 1938. *The Commercial Empire of the St Lawrence, 1760 -1850,* by D. G. Creighton, Toronto, 1937, lucidly describes the effects of canal development upon trading at Montreal.

The Canadian Canals: their history and cost, with An Inquiry into the policy necessary to advance the well-being of the Province is the Victorian title of a 191 page volume written by William Kingsford and published in Toronto in 1865. Although it is a piece of special pleading for the St Lawrence route to the Great Lakes, as against the Ottawa route, it does give a useful history of all the early canals. No trace has yet been found of any other book dealing with Canadian Canals as a whole even though, as will be seen from the *Notes,* there are a number of books on individual projects.

The Department of Transport published annually, until 1959, a pamphlet guide (of about 50 pages) to *The Canals of Canada,* giving factual information and brief historical resumés of all canals under its jurisdiction. Similar information is now available from the St Lawrence Seaway Authority and the Department of Indian Affairs and Northern Development but not in such convenient form.

Author's Acknowledgements

I am indebted to Mr David St John Thomas and Mr Charles Hadfield for the suggestion that this book should be written and for their continued encouragement. Mr Hadfield has been a most kindly and helpful editorial guide; I am grateful to him for valuable improvements in the text and for his care and attention to the final assembly of the supplementary material and illustrations.

Captain T. Appleton of Ottawa kindly read a draft of the complete text and favoured me with useful comments. Liaison has been maintained with Mr Harry Rinker of York, Pennsylvania so that his book on the canals of the United States of America may be a true companion volume.

In my studies of canals in the Maritime Provinces, I was kindly assisted by Messrs R.D. Evans, Carl Hudson, L. Keith Ingersoll, E.F. Osborne, D.C. Tibbetts and Drs Conrad and Esther Wright, Mr Lawrence Tombs of Montreal was good enough to share with me his long experience with the Chambly Canal, and Mr D.J. Gormley, the superintending engineer of the Trent Canal, assisted me with the capsule account of his extensive system.

Tracking down the long-forgotten details of some of the minor canals described in Chapter 6 was made the more pleasant through personal associations with many people, of whom I must mention with special appreciation Mr B.E. Marr, Mrs Kirstin Mueller, and Messrs S.B. Panting, J.C. Thatcher, G.B. Williams and Ian E. Wilson of Queen's University. After I had drafted a first note on the Baillie-

Grohman Canal, a description of this unusual venture was published in the *Bulletin* of the American Canal Society, to the good work of which I am glad to invite attention (PO Box 842, Shepherdstown, West Virginia, 25443).

Brigadier General L.C. Morrison, Director General of Information, Department of National Defence, Ottawa, and members of his staff went to more trouble than I had ever anticipated in obtaining for me confirmation of the local story about a destroyer in the St Peter's Canal and I am grateful. The statistics of the traffic through the canals of Canada that it has been possible to include in this volume in summary form demonstrate, almost without further comment, how fortunate Canada is in its national statistical service. Long known as the Dominion Bureau of Statistics but now as Statistics Canada, it is headed by Dr Sylvia Ostry as the Chief Statistician of Canada who kindly provided me with a copy of a most convenient summary paper, now long out-of-print.

Having long been a student of the St Lawrence Seaway, following its construction closely, I have been aided in these studies down the years by many friends, too numerous to mention here. I must, however, record my indebtedness to Captain K.R. Krabbe, Master of the M V *Havelland*, and his fellow officers, for their share in making a sail through the Seaway so memorable an experience.

In addition to some of those already mentioned, the ever-helpful staff of the Photo Section of the Public Archives of Canada assisted with the provision of illustrations from which those which now appear in this book were selected. In similarly helpful ways the staffs of the Notman Archives at McGill University, the Peterborough Public Library, Manitoba Government Travel, and Information Canada also provided illustrative material. Mrs Boone of the University of New Brunswick Library, Mrs Gay Hemsley of the St Lawrence Seaway Authority, Miss Margaret Van Every of the Archives of Ontario and Mrs Dalton of the Ontario Ministry of Industry and Tourism kindly gave their personal

attention and interest to my requests for assistance in supplementing the text with illustrations that would help readers to see something of what the canals of Canada look like.

Despite all this assistance, so willingly given, I am conscious of the impossibility of compressing into a volume of this size an adequate account of one of the great chapters in the history of engineering in Canada, now extending back for 180 years. For necessary omissions, for any lack of balance in what has been written and for such errors as are sure to have escaped the most careful checking possible, I accept full responsibility while hoping that this account, despite its imperfections, will at least show readers how greatly Canada is indebted to those who planned, designed, constructed and now operate her canals.

Ottawa; 24 xii 1974
R.F.L.

Index

PART TWO: PLACE NAMES

PART THREE: INDEX OF SUBJECTS